# CONTEMPORARY URBAN PLANNING

# CONTEMPORARY URBAN PLANNING

Robert Thomas

Kruger Brentt
Publishers
2023

Kruger Brentt Publishers UK. LTD.
Company Number 9728962

Regd. Office: 68 St Margarets Road, Edgware, Middlesex HA8 9UU

© 2023 AUTHOR
ISBN: 9781787150416

For information on all our publications visit our website at http://krugerbrentt.com/

# PREFACE

Urban planning is essential to the development of the communities around us – especially in and around cities and densely populated areas. It's a dynamic field that can look quite different by area of specialization. However, there's no question about just how important urban planners is to the development of societies as a whole. People in these roles are responsible for developing policies and systems that create the physical and economic infrastructure that supports diverse and growing populations. In doing this, they must effectively manage resources and plan for current and future needs of the population. If an urban area is badly planned, residents face traffic congestion, inadequate infrastructure, unaffordable housing, and are vulnerable to climate change, fire hazards, and flooding. The **city and its infrastructure become unsustainable** and ultimately hinder the growth of the population and economy. Urban planning also affects our national parks, **ecological diversity, heritage sites and natural resources** by ensuring that cities and towns are arranged so they don't encroach on these areas. By creating and enforcing laws to protect these unique Australian attractions, they will always be around for our children and tourists to appreciate.

While much of our urban planning has been reactionary, the goal is to enact planning regulations with a view for the future. By conducting censuses, governments can predict the changing demographics and growth of the population. Successful urban planning prepares for such changes by ensuring access to quality, affordable housing. Urban planning also needs to ensure ecological sustainability by prohibiting development that will destroy wildlife habitats such as those of our koala population. Environmental sustainability also refers to minimising carbon emissions by building cycle paths and promoting public transport by making it widely available. Well planned cities are also water sensitive, ensuring minimal water waste.

The present book entitled *"Contemporary Urban Planning"* contains eleventh chapters covering all related disciplines. These chapters include Urban Planning: An Introduction, Urban Planning and Urban Design, Importance of Urban Design,

Collaboration Between Architects and Planners in An Urban Design Studio, Urban Design, Socialization and Quality of Life, The Public Realm as the Armature of Civic Life, The Urban Planning Concept Based on Smart City Approach, Urban Planning Management, Integrative Theory Approach to Sustainable Urban Design: The Value of Geodesign, Role of Urban Design in developing Communities and Contemporary Urban Design: Theory and Practice Related terminology is given at the end for ready reference. It is an essential resource for students, city planners, and all who are concerned with the nature of contemporary urban development problems.

We are grateful to all those persons as well as various books, manuals, periodicals, magazines, journals etc. that helped in the preparation of this book. In spite of the best efforts, it is possible that some errors may have occurred into the compilation and editing of the book. Further queries, constructive suggestions and criticisms for the improvement of the book are always welcome and shall be thankfully acknowledged.

**Robert Thomas**

# CONTENTS

# 1

# URBAN PLANNING: AN INTRODUCTION

## 1.1 INTRODUCTION: AN OVERVIEW

Urban planning, design and regulation of the uses of space that focus on the physical form, economic functions, and social impacts of the urban environment and on the location of different activities within it. Because urban planning draws upon engineering, architectural, and social and political concerns, it is variously a technical profession, an endeavour involving political will and public participation, and an academic discipline. Urban planning concerns itself with both the development of open land ("greenfields sites") and the revitalization of

*Urban planning encompasses the preparation of plans for and the regulation and management of towns, cities, and metropolitan regions. It attempts to organize sociospatial relations across different scales of government and governance. Urban planning is concerned with the social, economic, and environmental consequences of delineating spatial boundaries and influencing spatial distributions of resources.*

existing parts of the city, thereby involving goal setting, data collection and analysis, forecasting, design, strategic thinking, and public consultation. Increasingly, the technology of geographic information systems (GIS) has been used to map the existing urban system and to project the consequences of changes. In the late 20th century the term sustainable development came to represent an ideal outcome in the sum of all planning goals. As advocated by the United Nations-sponsored World Commission on Environment and Development in Our Common Future (1987), sustainability refers to "development that meets the needs of the present without compromising the ability of future generations to meet their own needs." While there is widespread consensus on this general goal, most major planning decisions involve trade-offs between subsidiary objectives and thus frequently involve conflict.

The modern origins of urban planning lie in a social movement for urban reform that arose in the latter part of the 19th century as a reaction against the disorder of the industrial city. Many visionaries of the period sought an ideal city, yet practical considerations of adequate sanitation, movement of goods and people, and provision of amenities also drove the desire for planning. Contemporary planners seek to balance the conflicting demands of social equity, economic growth, environmental sensitivity, and aesthetic appeal. The result of the planning process may be a formal master plan for an entire city or metropolitan area, a neighbourhood plan, a project plan, or a set of policy alternatives. Successful implementation of a plan usually requires entrepreneurship and political astuteness on the part of planners and their sponsors, despite efforts to insulate planning from politics. While based in government, planning increasingly involves private-sector participation in "public-private partnerships."

Urban planning emerged as a scholarly discipline in the 1900s. In Great Britain the first academic planning program began at the University of Liverpool in 1909, and the first North American program was established at Harvard University in 1924. It is primarily taught at the postgraduate level, and its curriculum varies widely from one university to another. Some programs maintain the traditional emphasis on physical design and land use; others, especially those that grant doctoral degrees, are oriented toward the social sciences. The discipline's theoretical core, being somewhat amorphous, is better defined by the issues it addresses than by any dominant paradigm or prescriptive approach. Representative issues especially concern the recognition of a public interest and how it should be determined, the physical and social character of the ideal city, the possibility of achieving change in accordance with consciously determined goals, the extent to which consensus on goals is attainable through communication, the role of citizens versus public officials and private investors in shaping the city, and, on a methodological level, the appropriateness of quantitative analysis and the "rational model" of decision making. Most degree programs in urban planning consist principally of applied courses on topics ranging from environmental policy to transportation planning to housing and community economic development.

## 1.2 THE DEVELOPMENT OF URBAN PLANNING

### 1.2.1 Early history

Evidence of planning has been unearthed in the ruins of cities in China, India, Egypt, Asia Minor, the Mediterranean world, and South and Central America. Early examples of efforts toward planned urban development include orderly street systems that are rectilinear and sometimes radial; division of a city into specialized functional quarters; development of commanding central sites for palaces, temples, and civic buildings; and advanced systems of fortification, water supply, and drainage. Most of the evidence is in smaller cities that were built in comparatively short periods as colonies. Often the central cities of ancient states grew to substantial size before they achieved governments capable of imposing controls.

For several centuries during the Middle Ages, there was little building of cities in Europe. Eventually towns grew up as centres of church or feudal authority, of marketing or trade. As the urban population grew, the constriction caused by walls and fortifications led to overcrowding, the blocking out of air and light, and very poor sanitation. Certain quarters of the cities, either by custom or fiat, were restricted to different nationalities, classes, or trades, as still occurs in many contemporary cities of the developing world.

The physical form of medieval and Renaissance towns and cities followed the pattern of the village, spreading along a street or a crossroads in circular patterns or in irregular shapes, though rectangular patterns tended to characterize some of the newer towns. Most streets were little more than footpaths—more a medium for communication than for transportation—and even in major European cities paving was not widely introduced before the 12th century (1184 in Paris, 1235 in Florence, and 1300 in Lübeck). As the population of the city grew, walls were often expanded, but few cities at the time exceeded a mile in length. Sometimes sites were changed, as in Lübeck, and many new cities emerged with increasing population—frequently about one day's walk apart. Towns ranged in population from several hundred to perhaps 40,000 (as in London in the late 14th century, although London's population had been as high as 80,000 before the arrival of the Black Death). Paris and Venice were exceptions, reaching 100,000.

Conscious attempts to plan cities reemerged in Europe during the **Renaissance**. Although these efforts partly aimed at improving circulation and providing military defense, their prime objective was often the glorification of a ruler or a state. From the 16th century to the end of the 18th, many cities were laid out and built with monumental splendour. The result may have pleased and inspired the citizens, but it rarely contributed to their health, to the comfort of their homes, or to efficiency in manufacturing, distribution, and marketing.

The New World absorbed the planning concepts of European absolutism to only a limited degree. Pierre L'Enfant's grandiose plan for Washington, D.C. (1791), exemplified this transference, as did later City Beautiful projects, which aimed for grandeur in the siting of public buildings but exhibited less concern for the efficiency of residential, commercial, and industrial development. More influential on the layout of U.S. cities, however, was the rigid grid plan of Philadelphia, designed by **William Penn** (1682). This plan traveled west with the pioneers, since it was the simplest method of dividing surveyed territory. Although it took no cognizance of topography, it facilitated the development of land markets by establishing standard-sized lots that could be easily bought and sold—even sight unseen.

In much of the world, city plans were based on the concept of a centrally located public space. The plans differed, however, in their prescriptions for residential development. In the United States the New England town grew around a central commons; initially a pasture, it provided a focus of community life and a site for a meetinghouse, tavern, smithy, and shops and was later reproduced in the central squares of cities and towns throughout the country. Also from the New England town came the tradition of the freestanding single-family house that became the norm for most metropolitan areas. The central plaza, place, or square provided a focal point for European city plans as well. In contrast to American residential development, though, European domestic architecture was dominated by the attached house, while elsewhere in the world the marketplace or bazaar rather than an open space acted as the cynosure of cities. Courtyard-style domiciles characterized the Mediterranean region, while compounds of small houses fenced off from the street formed many African and Asian settlements.

## 1.2.2 The Era of Industrialization

In both Europe and the United States, the surge of industry during the mid- and late 19th century was accompanied by rapid population growth, unfettered business enterprise, great speculative profits, and public failures in managing the unwanted physical consequences of development. Giant sprawling cities developed during this era, exhibiting the luxuries of wealth and the meanness of poverty in sharp juxtaposition. Eventually the corruption and exploitation of the era gave rise to the Progressive movement, of which city planning formed a part. The slums, congestion, disorder, ugliness, and threat of disease provoked a reaction in which sanitation improvement was the first demand. Significant betterment of public health resulted from engineering improvements in water supply and sewerage, which were essential to the further growth of urban populations. Later in the century the first housing reform measures were enacted. The early regulatory laws (such as Great Britain's Public Health Act of 1848 and the New York State Tenement House Act of 1879) set minimal standards for housing construction. Implementation, however, occurred only slowly, as governments did not provide funding for upgrading existing dwellings, nor did the minimal rent-paying ability of slum dwellers offer incentives for landlords to improve their buildings. Nevertheless, housing improvement occurred as new structures were erected, and new legislation continued to raise standards, often in response to the exposés of investigators and activists such as Jacob Riis in the United States and Charles Booth in England.

Also during the Progressive era, which extended through the early 20th century, efforts to improve the urban environment emerged from recognition of the need for recreation. Parks were developed to provide visual relief and places for healthful play or relaxation. Later, playgrounds were carved out in congested areas, and facilities for games and sports were established not only for children but also for adults, whose workdays gradually shortened. Supporters of the parks movement believed that the opportunity for outdoor recreation would have a civilizing effect on the working classes, who were otherwise consigned to overcrowded housing and unhealthful workplaces. New York's Central Park, envisioned in the 1850s and designed by architects Calvert Vaux and Frederick Law Olmsted, became a widely imitated model. Among its contributions were the separation of pedestrian and vehicular traffic, the creation of a romantic landscape within the heart of the city, and a demonstration that the creation of parks could greatly enhance real-estate values in their surroundings.

Concern for the appearance of the city had long been manifest in Europe, in the imperial tradition of court and palace and in the central plazas and great buildings of church and state. In Paris during the Second Empire (1852–70), Georges-Eugène, Baron Haussmann, became the greatest of the planners on a grand scale, advocating straight arterial boulevards, advantageous vistas, and a symmetry of squares and radiating roads. The resulting urban form was widely emulated throughout the rest of continental Europe. Haussmann's efforts went well beyond beautification, however; essentially they broke down the barriers to commerce presented by medieval Paris, modernizing the city so as to enable the efficient transportation of goods as well as the rapid mobilization of military troops. His designs involved the demolition of antiquated tenement structures and their replacement by new apartment houses intended for a wealthier clientele, the construction of transportation corridors and commercial space that broke up residential neighbourhoods, and the displacement of poor people from centrally located areas. Haussmann's methods provided a template by which urban redevelopment programs would operate in Europe and the United States until nearly the end of the 20th century, and they would extend their influence in much of the developing world after that.

As the grandeur of the European vision took root in the United States through the City Beautiful movement, its showpiece became the World's Columbian Exposition of 1893, developed in Chicago according to principles set out by American architect Daniel Burnham. The architectural style of the exposition established an ideal that many cities imitated. Thus, the archetype of the City Beautiful—characterized by grand malls and majestically sited civic buildings in Greco-Roman architecture—was replicated in civic centres and boulevards throughout the country, contrasting with and in protest against the surrounding disorder and ugliness. However, diffusion of the model in the United States was limited by the much more restricted power of the state (in contrast to European counterparts) and by the City Beautiful model's weak potential for enhancing businesses' profitability.

Whereas Haussmann's approach was especially influential on the European continent and in the design of American civic centres, it was the utopian concept of the garden city, first described by British social reformer Ebenezer Howard in his book Garden Cities of To-Morrow (1902), that shaped the appearance of residential areas in the United States and Great Britain. Essentially a suburban form, Howard's garden city incorporated low-rise homes on winding streets and culs-de-sac, the separation of commerce from residences, and plentiful open space lush with greenery. Howard called for a "cooperative commonwealth" in which rises in property values would be shared by the community, open land would be communally held, and manufacturing and retail establishments would be clustered within a short distance of residences. Successors abandoned Howard's socialist ideals but held on to the residential design form established in the two new towns built during Howard's lifetime (Letchworth and Welwyn Garden City), ultimately imitating the garden city model of winding roads and ample greenery in the forming of the modern suburban subdivision.

Perhaps the single most influential factor in shaping the physical form of the contemporary city was transportation technology. The evolution of transport modes from foot and horse to mechanized vehicles facilitated tremendous urban territorial expansion. Workers were able to live far from their jobs, and goods could move quickly from point of production to market. However, automobiles and buses rapidly congested the streets in the older parts of cities. By threatening strangulation of traffic, they dramatized the need to establish new kinds of orderly circulation systems. Increasingly, transportation networks became the focus of planning activities, especially as subway systems were constructed in New York, London, and Paris at the beginning of the 20th century. To accommodate increased traffic, municipalities invested heavily in widening and extending roads. (See also traffic control).

Many city governments established planning departments during the first third of the 20th century. The year 1909 was a milestone in the establishment of urban planning as a modern governmental function: it saw the passage of Britain's first town-planning act and, in the United States, the first national conference on city planning, the publication of Burnham's plan for Chicago, and the appointment of Chicago's Plan Commission (the first recognized planning agency in the United States, however, was created in Hartford, Connecticut, in 1907). Germany, Sweden, and other European countries also developed planning administration and law at this time.

The colonial powers transported European concepts of city planning to the cities of the developing world. The result was often a new city planned according to Western principles of beauty and separation of uses, adjacent to unplanned settlements both new and old, subject to all the ills of the medieval European city. New Delhi, India, epitomizes this form of development. Built according to the scheme devised by the British planners Edwin Lutyens and Herbert Baker, it grew up cheek by jowl with the tangled streets of Old Delhi. At the same time, the old city, while less salubrious, offered its inhabitants a sense of community, historical continuity, and functionality more suited to their way of life. The same pattern repeated itself throughout the British-ruled territories, where African capitals such as Nairobi, Kenya, and Salisbury, Southern Rhodesia (now Harare, Zimbabwe), were similarly designed to accommodate their white colonial rulers. Although the decorative motifs imposed by France in its colonial capitals reflected a somewhat different aesthetic sensibility, French planners likewise implanted broad boulevards and European-style housing in their colonial outposts.

# 1.3 URBAN FORM

## 1.3.1 Zoning and Subdivision Controls

As Western industrial cities rapidly expanded during the first part of the 20th century, factories encroached upon residential areas, tenements crowded in among small houses, and skyscrapers overshadowed other buildings. To preserve property values and achieve economy and efficiency in the structure and arrangement of the city, policy makers perceived a need to sort out incompatible activities, set some limits upon building size, and protect established areas from despoilment. Master plans prescribed the desired patterns of traffic circulation, bulk and density levels, and necessary public improvements. Zoning regulations, first instituted in the early decades of the 20th century, were the principal means for achieving these goals. They set maximums for building breadth and height and designated acceptable configurations of structures within demarcated areas (zones); most important in terms of their effect on urban development, zoning codes segregated particular uses of urban space. Thus, housing, manufacturing, and retail activities, which formerly intermixed, now took place in different parts of the city. Although zoning protected residents from adjacent noxious uses, it had the less-desirable further effect of forcing long trips to work and increasing routine travel, thereby contributing to traffic congestion and limiting activity in each part of the city to different times of the day. Some zoning codes provoked disputes. Court cases in the United States challenged zoning ordinances that, by requiring large single-residence dwellings on large lots, restricted the construction of affordable homes for low-income households. In some states courts struck down exclusionary zoning, and some remedial legislation was passed.

Parallel to the evolution of zoning in the United States was the development of subdivision controls, which subjected the initial laying out of vacant land to public regulation. These regulations affected the design of new developments and specified that new streets had to conform to the overall city plan. Some subdivision ordinances required property developers to provide the land needed for streets, playgrounds, and school sites and to pay all or most of the cost of building these facilities.

## 1.3.2 New towns

After World War II a number of European countries, especially France, the Netherlands, Germany, and the Soviet Union, undertook the building of new towns (comprehensive new developments outside city centres) as governmental enterprises. Concerned with what they regarded as too much density within urban areas, governments constructed these new towns as a means of capturing the overspill from cities within planned developments rather than allowing haphazard exurban growth. Most of them, except in the Soviet Union, were primarily residential suburbs, although some British towns such as **Milton Keynes** did succeed in attracting both industry and population within low-rise conurbations. In Sweden the government successfully constructed accessible high-rise residential suburbs with mixed-income occupancy. Tapiola, in metropolitan Helsinki, Finland, was a low-rise ensemble embodying many of Howard's original ideas and incorporating **architecture** of the highest order. New town development in France, Italy, Spain, and Belgium, however, mostly resulted in large, uninviting high-rise residential projects for the working class on the urban **periphery**.

American postwar new town development depended largely on private **initiative**, with Reston, Virginia; Columbia, Maryland; Irvine, California; and Seaside, **Florida**, serving as some of the better-known examples. Preceding these efforts, however, were a number of small, privately planned suburbs, including Riverside, **Illinois**, a planned **community** outside **Chicago** that was designed by **Frederick Law Olmsted** in 1868–69, and Radburn, **New Jersey**, built in 1929 according to plans conceived by Clarence Stein and Henry Wright. There are a few outstanding examples of planned new cities in such widely scattered places as India (where **Le Corbusier** designed **Chandigarh**), the **Middle East**, and **South America**.

In **Asia** the emerging industrial economies of the post-World War II period produced large, densely populated, congested metropolises. Some Asian governments addressed the problems of rapid expansion through massive construction projects that **encompassed** skyscraper office buildings, shopping malls, luxury apartments and hotels, and new airports. In Shanghai, in the span of little more than a decade, the Chinese government created Pudong New Area—a planned central business district along with factories and residences in Pudong, across the Huangpu River from Shanghai's old downtown core. Many developing countries, however, are still preoccupied with political and economic problems and have made little progress toward establishing an environmental planning function capable of avoiding the insalubrious conditions that characterized Western cities in the 19th century.

## 1.4 THE SCOPE OF PLANNING

Throughout the first half of the 20th century, the influence of planning broadened within Europe as various national and local statutes increasingly guided new development. European governments became directly involved with housing provision for the working class, and decisions concerning the siting of housing construction shaped urban growth. In the United States, local planning in the form of zoning began with the 1916 New York City zoning law, but it was not until the Great Depression of the 1930s that the federal government intervened in matters of housing and land use. During World War II, military mobilization and the need to coordinate defense production caused the development of the most extensive planning frameworks ever seen in the United States and Britain. Although the wartime agencies were demobilized after hostilities ended, they set a precedent for national economic and demographic planning, which, however, was much more extensive in Britain than in the United States.

### 1.4.1 Postwar approaches

During the postwar period European governments mounted massive housing and rebuilding programs within their devastated cities. These programs were guided by the principles of modernist planning promulgated through the Congrès International d'Architecture Moderne (CIAM), based on the ideas of art and architectural historian Siegfried Giedion, Swiss architect Le Corbusier, and the International school rooted in Germany's Bauhaus. High-rise structures separated by green spaces prevailed in the developments built during this period. Their form reflected both the need to produce large-scale, relatively inexpensive projects and the architects' preference for models that exploited new materials and technologies and could be replicated universally. Government involvement in housing development gave the public sector a more direct means of controlling the pattern of urban growth through its investments, rather than relying on regulatory devices as a means of restricting private developers.

Within Britain the Greater London Plan of Leslie Patrick Abercrombie called for surrounding the metropolitan area with an inviolate greenbelt, construction of new towns beyond the greenbelt that would allow for lowering of population densities in the inner city, and the building of circumferential highways to divert traffic from the core. The concept of the sharp separation of city from country prevailed also throughout the rest of Britain and was widely adopted in the Scandinavian countries, Germany, and the Netherlands as well. In the United States the burgeoning demand for housing stimulated the construction of huge suburban subdivisions. Construction was privately planned and financed, but the federal government encouraged it through tax relief for homeowners and government-guaranteed mortgages. Suburban planning took place at the municipal level in the form of zoning and subdivision approval, public development of sewerage and water systems, and schools. The lack of metropolitan-wide planning jurisdictions resulted in largely unplanned growth and consequent urban sprawl. Within central cities, however, the federal government subsidized land clearance by local urban renewal authorities and the construction of public housing (i.e., publicly owned housing for low-income people). Local government restricted its own reconstruction activities to public facilities such as schools, police stations, and recreation centres. It relied on private investors for the bulk of new construction, simply indicating what would be desirable. Consequently, many cleared sites lay vacant for decades when the private market did not respond.

### 1.4.2 Planning and Government

The place of the city-planning function in the structure of urban government developed in different ways in different countries. In many countries today, private developers must obtain governmental permission in order to build. In the United States, however, they may build "as of right" if their plans conform to the municipality's zoning code. On the European continent, where municipal administration is strongly centralized, city planning occurs within the sphere of an executive department with substantial authority.

In the United Kingdom the local planning authority is the elected local council, while a planning department acts in an executive and advisory capacity. DeveAlthough the mayor and council have final decision-making power in U.S. cities, an independent planning commission of appointed members usually takes primary responsibility for routine planning functions. Planning activity primarily consists of the approval or disapproval of private development proposals. In larger cities the commission has a staff reporting to it. During the period of a large, federally financed urban renewal program in place from 1949 to 1974, most American cities had powerful semi-independent urban renewal authorities that were responsible for redevelopment planning. Some of these still exist, but in most places they either became subordinate to the mayor or combined with economic development agencies, which are often quasi-autonomous corporations. While they are appointed by the mayor and council, these agencies usually report to an independent board of directors drawn primarily from the business community. Especially as city government became preoccupied with economic development planning, the agencies were authorized to enter into development agreements with private investors.

In some countries, most notably in northern Europe, national governments made city planning part of their overall effort to deal with issues of growth and social welfare. Even in the United States—where the initiative remained with local governments and where metropolitan government never gained a significant foothold—the federal government became involved with local planning issues through the creation and execution of national housing and urban renewal legislation and through the supervisory role of the federal Department of Housing and Urban Development, established in 1965. As developing countries gained independence from colonial powers in the 1960s and '70s, planning structures became highly centralized within the new national governments, which typically laid down the framework for city planning.

### 1.4.3 Competing Models

Starting in the 20th century, a number of urban planning theories came into prominence and, depending on their popularity and longevity, influenced the appearance and experience of the urban landscape. The primary goal of city planning in the mid-20th century was comprehensiveness. An increasing recognition of the interdependence of various aspects of the city led to the realization that land use, transport, and housing needed to be designed in relation to each other. Developments in other disciplines, particularly management science and operations research, influenced academic planners who sought to elaborate a universal method—also known as "the rational model"—whereby experts would evaluate alternatives in relation to a specified set of goals and then choose the optimum solution. The rational model was briefly hegemonic, but this scientific approach to public-policy making was quickly challenged by critics who argued that the human consequences of planning decisions could not be neatly quantified and added up.

The modernist model, involving wholesale demolition and reconstruction under the direction of planning officials isolated from public opinion, came under fierce attack both intellectually and on the ground. Most important in undermining support for the modernist approach was urbanologist Jane Jacobs. In her book The Death and Life of Great American Cities (1961), she sarcastically described redeveloped downtowns and housing projects as comprising the "radiant garden city"—a sly reference to the influence of Le Corbusier's "towers in the park" and Ebenezer Howard's antiurban garden city. Jacobs criticized large-scale clearance operations for destroying the complex social fabric of cities and imposing an inhuman orderliness. Rather than seeing high population density as an evil, she regarded it as an important factor in urban vitality. She considered that a lively street life made cities attractive, and she promoted diversity of uses and population groups as a principal value in governing urban development. According to Jacobs, urban diversity contributes to sustainable growth, whereas undifferentiated urban settings tend to depend upon unsustainable exploitation, exhibited in the extreme form by lumber or mining towns that collapse after the valuable resources have been removed. Jacobs was not alone in her criticism. Beginning in the 1960s, urban social movements, at times amounting to insurrection, opposed the displacements caused by large-scale modernist planning. In cities throughout the United States and Europe, efforts at demolishing occupied housing provoked fierce opposition. Within developing countries, governmental attempts to destroy squatter settlements stimulated similar counteroffensives.

By the end of the 20th century, planning orthodoxy in the United States and Europe began to take Jacobs's arguments into account. New emphasis was placed on the rehabilitation of existing buildings, historical preservation, adaptive reuse of obsolete structures, mixed-use development, and the "24-hour city"—i.e., districts where a variety of functions would create around-the-clock activity. Major new projects, while still sometimes involving demolition of occupied housing or commercial structures, increasingly came to be built on vacant or "brownfields" sites such as disused railroad yards, outmoded port facilities, and abandoned factory districts. Within developing countries, however, the modernist concepts of the earlier period still retained a significant hold. Thus, for example, China, in preparation for the Beijing Olympics of 2008, engaged in major displacement of its urban population to construct roads and sports facilities, and it likewise developed new commercial districts by building high-rise structures along the functionalist Corbusian model.

### 1.4.4 Contemporary Planning

The ways in which planning operated at the beginning of the 21st century did not conform to a single model of either a replicable process or a desirable outcome. Within Europe and the United States, calls for a participatory mode—one that involved residents most likely to be affected by change in the planning process for their locales—came to be honoured in some cities but not in others. The concept of participatory planning has spread to the rest of the world, although it remains limited in its adoption. Generally, the extent to which planning involves public participation reflects the degree of **democracy** enjoyed in each location. Where government is **authoritarian**, so is planning. Within a more participatory framework, the role of planner changes from that of expert to that of mediator between different groups, or "stakeholders." This changed role has been **endorsed** by theorists supporting a concept of "communicative rationality." Critics of this viewpoint, however, argue that the process may suppress **innovation** or simply promote the wishes of those who have the most power, resulting in outcomes contrary to the public interest. They are also concerned that the response of "not in my backyard" ("NIMBYism") precludes building **affordable housing** and needed public facilities if neighborhood residents are able to veto any construction that they fear will lower their property values.

In sum, the enormous variety of types of projects on which planners work, the lack of consensus over processes and goals, and the varying approaches taken in different cities and countries have produced great variation within contemporary urban planning. Nevertheless, although the original principle of strict segregation of uses continues to prevail in many places, there is an observable trend toward mixed-use development— particularly of complementary activities such as retail, entertainment, and housing— within urban centres.

### 1.4.5 Changing Objectives

Although certain goals of planning, such as protection of the environment, remain important, emphases among the various objectives have changed. In particular, economic development planning, especially in old cities that have suffered from the decline of manufacturing, has come to the fore. Planners responsible for economic development behave much like business executives engaged in marketing: they promote their cities to potential investors and evaluate physical development in terms of its attractiveness to capital and its potential to create jobs, rather than by its healthfulness or conformity to a master plan. Such planners work to achieve development agreements with builders and firms that will contribute to local commerce. Especially in the United States and the United Kingdom, planning agencies have concerned themselves with promoting economic development and have become involved in negotiating deals with private developers. In the United Kingdom these can include the trading of planning permission for "planning gain" or other community benefits; in other words, developers may be allowed to build in return for providing funds, facilities, or other benefits to the community. In the United States, where special permission is not required if the building fits into the zoning ordinance, deals usually involve some kind of public subsidy. Typical development agreements involve offering land, tax forgiveness, or regulatory relief to property developers in return for a commitment to invest in an area or to provide amenities. An agreement may also be struck between the city and a private firm in which the firm agrees to move into or remain in an area in return for various concessions. Many such arrangements generate controversy, especially if a municipality exercises the right of eminent domain and takes privately owned land for development projects.

A late 20th-century movement in planning variously called new urbanism, smart growth, or neotraditionalism, has attracted popular attention through its alternative views of suburban development. Reflecting considerable revulsion against urban sprawl, suburban traffic congestion, and long commuting times, this movement has endorsed new construction that brings home, work, and shopping into proximity, encourages pedestrian traffic, promotes development around mass-transit nodes, and mixes types of housing. Within the United Kingdom, Prince Charles became a strong proponent of neotraditional planning through his sponsorship of Poundbury, a new town of traditional appearance in Dorset. Similar efforts in the United States, where growth on the metropolitan periphery continued unabated, chiefly arose as limited areas of planned development amid ongoing dispersal and sprawl. Although the movement's primary influence has been in new suburban development, it has also been applied to the redevelopment of older areas within the United Kingdom and the United States. Paternoster Square in London, adjacent to Saint Paul's Cathedral, and a number of HOPE VI schemes in the United States (built under a federal program that demolished public housing projects and replaced them with mixed-income developments) have been erected in accordance with neotraditional or new urbanist ideas.

### 1.4.6 New Pluralism

Universal principles regarding appropriate planning have increasingly broken down as a consequence of several trends. First, intellectual arguments against a "one plan fits all" approach have gained ascendancy. The original consensus on the form of orderly development embodying separation of uses and standardized construction along modernist lines has been replaced by sensitivity to local differences and greater willingness to accept democratic input. Second, it has become widely recognized that, even where the imposition of standards might be desirable, many places lack the resources to attain them. Within the developing world, informal markets and settlements, formerly condemned by planners, now appear to be inevitable and often appropriate in serving the needs of poor communities. Planners in these contexts, influenced by international aid institutions, increasingly endeavour to upgrade squatter settlements and street markets rather than eliminate them in the name of progress. Third, political forces espousing the free market have forced planners to seek market-based solutions to problems such as pollution and the provision of public services. This has led to privatization of formerly publicly owned facilities and utilities and to the trading of rights to develop land and to emit pollutants in place of a purely regulatory approach.

Planning in its origins had an implicit premise that a well-designed, comprehensively planned city would be a socially ameliorative one. In other words, it tended toward environmental determinism. The goals of planning have subsequently become more modest, and the belief that the physical environment can profoundly affect social behaviour has diminished. Nevertheless, planning as practice and discipline relies upon public policy as an instrument for producing a more equitable and attractive environment that, while not radically altering human behaviour, nonetheless contributes to improvements in the quality of life for a great number of people.

# 1.5 STRATEGIC PLANNING

## 1.5.1 Organizational Management

Strategic planning, disciplined effort to produce decisions and actions that shape and guide an organization's purpose and activities, particularly with regard to the future. Strategic planning is a fundamental component of organizational management and decision making in public, private, and nonprofit organizations. It is a structured approach to establishing an organization's direction and to anticipating the future. Through strategic planning, resources are concentrated on a limited number of objectives, thereby helping an organization to focus its efforts, to ensure that its members are working toward the same goals, and to assess and adjust its direction in response to a changing environment.

The process of strategic planning is disciplined in that it raises a sequence of questions that helps organizational leadership examine experience, test assumptions, gather and incorporate information about the present, and anticipate the environment in which the organization will be working in the future. By setting priorities, strategic planning implies that some organizational decisions and actions are more important than others. Much of the strategy lies in making difficult decisions about what is most important to achieving organizational effectiveness. Typically, the strategy encompasses activity over several years and needs to be altered over the course of time.

There are a variety of perspectives, models, and approaches used in strategic planning. The way that a strategic plan is developed depends on the nature of the organization's leadership, the culture of the organization, the complexity of the organization and its environment, and the size of the organization.

## 1.5.2 The evolution of Strategic Planning

The modern concept of corporate strategic planning grew out of budget exercises carried out in the 1950s in the United States. By the mid-1960s and throughout the 1970s, strategic planning was occurring in most large corporations. During this time, the U.S. government introduced program budgeting as a way of recording detailed information on costs associated with specific activities covered by a budget. Public and nonprofit organizations recognized the usefulness of strategy formulation during the 1980s, when the notion of marketing for public and nonprofit organizations gained prominence. Most well-known models of public and nonprofit strategic planning have their roots in the Harvard policy model developed at the Harvard Business School at Harvard University in the United States. The systematic analysis of strengths, weaknesses, opportunities, and threats (SWOT) is a primary strength of the Harvard model and constitutes a step in the strategic-planning model.

## 1.6 BENEFITS OF STRATEGIC PLANNING

Strategic planning clearly defines the purpose of the organization and establishes realistic goals and objectives consistent with that mission in a defined time frame within the organization's capacity for implementation. It communicates those goals and objectives to the organization's constituents. Strategic planning develops a sense of ownership of the plan and ensures that the most-effective use is made of the organization's resources by focusing the resources on key priorities. It provides a base from which progress can be measured and establishes a mechanism for informed change when needed.

The indicators to be used in assessing organizational effectiveness must be chosen from several possible areas and data gathered from several possible sampling frames. The pattern of strategy in an organization is determined not only by the plans and actions of its leaders but also by forces in its external environment. Because both organizations and environments can change over time, and because different agencies operate under different conditions, no single strategy is universally viable.

Organizations cannot be effective unless they know where they are headed. Effectiveness is not random; it begins with a clear vision, mission, and goals. Formal strategic-planning approaches establish those missions, goals, and visions. Strategic management offers a means of systematically thinking about and reviewing an organization's direction, environment, and strategies. Strategic planning is essential and continues the process for public organizations that wish to determine their own vision and mission. But strategic planning and continuous change require committed leadership, a supportive organizational culture, an established structure for coordinating and managing the implementation process, and the ability on the part of organizational members to participate in the planning process. Participation can be a powerful device for directing the energy of participants in the public organization.

## 1.7 ARCHITECTS AND ARTISTS

### 1.7.1 Inigo Jones (English Architect and Artist)

Inigo Jones, (born July 15, 1573, Smithfield, London, Eng.—died June 21, 1652, London), British painter, architect, and designer who founded the English classical tradition of architecture. The Queen's House (1616–19) at Greenwich, London, his first major work, became a part of the National Maritime Museum in 1937. His greatest achievement is the Banqueting House (1619–22) at Whitehall. Jones's only other surviving royal building is the Queen's Chapel (1623–27) at St. James's Palace. Jones was the son of a cloth worker also called Inigo. Of the architect's early life little is recorded, but he was probably apprenticed to a joiner. By 1603 he had visited Italy long enough to acquire skill in painting and design and to attract the patronage of King Christian IV of Denmark and Norway, at whose court he was employed for a time before returning to England. There he is next heard of as a "picture maker" (easel painter). Christian IV's sister, Anne, was the queen of James I of England, a fact that may have led to Jones's employment by her in 1605 to design the scenes and costumes of a masque, the first of a long series he designed for her and later for the king. The words to these masques were often supplied by Ben Jonson, the scenery, costumes, and effects nearly always by Jones. More than 450 drawings by him, representing work on 25 masques, a pastoral, and two plays ranging in date between 1605 and 1641, survive at Chatsworth House, Derbyshire.

From 1605 until 1610 Jones probably regarded himself as primarily under the queen's protection, but he was patronized also by Robert Cecil, 1st earl of Salisbury, for whom he produced his earliest known architectural work, a design for the New Exchange in the Strand (c. 1608; demolished in the 18th century). Though a somewhat immature design, the work was more sophisticated than anything being done in England at the time. Some designs (later superseded) for the restoration and improvement of Old St. Paul's Cathedral also date from this period, and in 1610 Jones was given an appointment that confirmed the direction of his future career. He became surveyor of works to the heir to the throne, Henry, prince of Wales.

This appointment, with all its promise, was short-lived, and Jones did little or nothing for the prince before the latter's death in 1612. In 1613, however, he was compensated by the guarantee of still higher office on the death of the king's surveyor of works, Simon Basil. To this office Jones succeeded in 1615, in the meantime having taken the opportunity offered him by Thomas Howard, 2nd earl of Arundel, to revisit Italy. Arundel and his party, including Jones, left England in April 1613 and proceeded to Italy, spending the winter of 14–1613 in Rome. In the course of the visit Jones had ample opportunity to study works by modern masters as well as antique ruins. Of the masters, the one to whom he attached the greatest importance was Andrea Palladio, the Italian architect who had gained wide influence through his The Four Books of Architecture (1570; I quattro libri dell'architettura), which Jones took with him on his tour. Returning to England in the autumn of 1614, Jones had completed his self-education as a classical architect.

Jones's career as surveyor of works to James I and Charles I lasted from 1615 to 1643. During most of those 28 years he was continuously employed in the building, rebuilding, or improvement of royal houses. His first important undertaking was the Queen's House at Greenwich, based to some extent on the Medici villa at Poggio a Caiano, near Florence, but detailed in a style closer to Palladio or Vincenzo Scamozzi (1616–1552). Work there was suspended on the death of Queen Anne in 1619 and completed only in 1635 for Charles's queen, Henrietta Maria. The building, considerably altered, now houses part of the National Maritime Museum.

In 1619 the Banqueting House at Whitehall was destroyed by fire; and between that year and 1622 Jones replaced it with what has always been regarded as his greatest achievement. The Banqueting House consists of one great chamber, raised on a vaulted basement. It was conceived internally as a basilica on the Vitruvian model but without aisles, the superimposed columns being set against the walls, which support a flat, beamed ceiling. For the main panels of this ceiling, allegorical paintings by Peter Paul Rubens were commissioned by Charles I and set in place in 1635. The exterior echoes the arrangement of the interior, with pilasters and regular columns set against rusticated walling.

The Banqueting House has only two complete facades. The ends were never completed, and this has given rise to the supposition that the building was intended to form part of a larger whole. This may have been so, and it is certain that Charles I, nearly 20 years after the Banqueting House was built, instructed Jones to prepare designs for rebuilding the whole of Whitehall Palace. These designs exist (at Worcester College, Oxford, and at Chatsworth House) and are among Jones's most interesting creations. They owe something to the palace of El Escorial near Madrid but are worked out in terms deriving partly from Palladio and Scamozzi and partly from Jones's own studies of the antique.

Jones's work was not confined to royal palaces. He was much involved in the regulation of new buildings in London, and out of this activity emerged the project that he planned in 1630 for the 4th earl of Bedford on his land at Covent Garden. This comprised a large open space bounded on the north and east by arcaded houses, on the south by the earl's garden wall, and on the west by a church with flanking gateways connecting to two single houses. The design probably derives partly from the piazza in Livorno, Italy, and partly from the Place Royale (now the Place des Vosges) in Paris. None of the original houses survive, but the church of St. Paul still stands, though much altered. Its portico is an instance, unique in Europe at its date of construction, of the use of the primitive Tuscan order of architecture.

With Covent Garden, Jones introduced formal town planning to London—it is the first London "square." He was probably instrumental, from 1638, in creating another square by planning the layout of the houses in Lincoln's Inn Fields, one of the houses (Lindsey House, still existing at No. 59 and 60) being attributed to him.

The most important undertaking of Jones's later years in office was the restoration of Old St. Paul's Cathedral in 42–1633. This included not only the repair of the 14th-century choir but the entire recasing, in rusticated masonry, of the Romanesque nave and transepts and the building of a new west front with a portico (56 feet [17 metres] high) of 10 columns. This portico, among Jones's most ambitious and subtly calculated works, tragically vanished with the rebuilding of the cathedral after the Great Fire of London in 1666. (In 1997 more than 70 carved stones from the portico were excavated from the building's foundations.) Jones's work at St. Paul's considerably influenced Sir Christopher

Wren and is reflected in some of his city churches as well as in his early designs for rebuilding the cathedral.

At the outbreak of the English Civil Wars in 1642, Jones was compelled to relinquish his office as surveyor of works and left London. He was captured at the siege of Basing House in 1645. His estate was temporarily confiscated, and he was heavily fined. In the following year, however, his pardon was confirmed by the House of Lords and his estate restored. In the year of Charles I's execution, 1649, he was doing work at Wilton for the earl of Pembroke, but the great double-cube room there is probably mostly the work of his pupil John Webb, who survived to reestablish something of the Jones tradition after the Restoration in 1660. Jones was buried with his parents in the church of St. Benet, Paul's Wharf, in London.

### 1.7.2 Pierre Charles L'Enfant (French engineer and architect)

Pierre Charles L'Enfant, (born August 2, 1754, Paris, France—died June 14, 1825, Prince George's county, Maryland, U.S.), French-born American engineer, architect, and urban designer who designed the basic plan for Washington, D.C., the capital city of the United States. L'Enfant studied art under his father at the Royal Academy of Painting and Sculpture from 1771 until he enlisted in 1776 as a volunteer in the American Continental Army. In recognition of his services, Congress made him major of engineers in 1783. The medal and diploma of the Society of the Cincinnati, an association of former Revolutionary officers, were designed by L'Enfant, and upon returning to Paris he helped organize the French branch of the society. L'Enfant went again to America in 1784 and settled in New York City. There, in addition to small architectural jobs, he renovated the old city hall for the U.S. Congress as Federal Hall (1788–89). For this, his first major architectural essay, he added star decorations to the Doric order in honour of his adopted country. He also designed the grandiose Morris House in Philadelphia, a mansard-styled structure that was begun in 1794 but was never completed.

When Congress decided to build a federal capital on the Potomac River, President George Washington hired L'Enfant in 1791 to prepare a plan for it. The plan he created was a gridiron of irregular rectangular blocks upon which broad diagonal avenues were superimposed. It was devised to focus on the Capitol and the presidential mansion and to form many squares, circles, and triangles at street intersections where monuments and fountains could be placed. The plan used to advantage the uneven ground and prepared for future transportation needs as well. Secretary of State Thomas Jefferson had provided L'Enfant with maps of various European cities to use as models, but, instead of copying any one of them, L'Enfant took ideas from several. The influence of Baroque planning at Versailles by André Le Nôtre appears in his plan, and it also bears resemblances to the London plans of Sir Christopher Wren and John Evelyn. Washington was forced to dismiss L'Enfant in 1792 for his obstinacy in defying the commissioners of the city, and particularly for his high-handed procedure in removing the house of Daniel

Carroll, an influential Washington resident, to make way for an avenue. Nevertheless, his plan of the city was generally followed. L'Enfant later attempted to obtain 95,500$ as payment for his services. Congress gave him what it thought to be proper, the sum of about 3,800$. In his old age L'Enfant lived with friends at Green Hill, a Maryland estate, where he died penniless. In 1909 his body was removed to Arlington National Cemetery, where a suitable monument was erected to him by Congress.

### 1.7.3 Jules Hardouin-Mansart (French architect)

Jules Hardouin-Mansart, (born c. April 16, 1646, Paris, France—died May 11, 1708, Marly-le-Roi), French architect and city planner to King Louis XIV who completed the design of Versailles. In 1675 Mansart became official architect to the king and from 1678 was occupied with redesigning and enlarging the palace of Versailles. He directed a legion of collaborators and protégés, many of whom became the leading architects of the following age. Starting from plans of architect Louis Le Vau, Mansart built the new Hall of Mirrors, the Orangerie, the Grand Trianon, and the north and south wings. At the time of his death he was working on the chapel. The vast complex, with an exquisite expanse of gardens designed by André Le Nôtre, was a harmonious expression of French Baroque classicism and a model that other courts of Europe sought to emulate. Although occupied with this enormous project for much of his life, Mansart built many other public buildings, churches, and sumptuous houses. Thought to be most reflective of his individual ability to combine classical and Baroque architectural design is the chapel of Les Invalides, Paris. Admirable contributions to city planning include his Place Vendôme and Place des Victoires, Paris.

### 1.7.4 Maxwell Fry (British architect)

Maxwell Fry, in full Edwin Maxwell Fry, (born Aug. 2, 1899, Wallasey, Cheshire, Eng.—died Sept. 3, 1987, Cotherstone, Durham), British architect who, with his wife, Jane Drew, pioneered in the field of modern tropical building and town planning. One of the earliest British adherents to the modern movement, Fry was trained at the School of Architecture, University of Liverpool. In 1924 he joined the town-planning firm of Adams and Thompson in London. Renouncing Classical architecture, he wrote that he saw "no place for it in a technocratic world." His early work shows the strong influence of Ludwig Mies van der Rohe, a leading proponent of the International style in architecture. In 1946 Fry and Drew (married 1942) formed the firm of Fry, Drew and Partners, London, specializing in large-scale planning for tropical countries. Among the many tropical buildings they designed are those of the University of Ibadan (1953–59), Nigeria. Their books Village Housing in the Tropics (1947; with Harry L. Ford) and Tropical Architecture in the Humid Zone (1956) are considered standard works. The Swiss-born architect Le Corbusier invited Fry and Drew to join him in 1951 on the project to build Chandigarh, the new capital city of the state of Punjab (from 1966 joint capital of Punjab and Haryana) in India. In their houses there Fry and Drew employed canopies and deep recesses for sun-sheltering purposes.

Fry's other important written works are The Bauhaus and the Modern Movement (1968) and Art in a Machine Age (1969).

### 1.7.5 Tatiana Bilbao (Mexican architect)

Tatiana Bilbao, (born August 2, 1972, Mexico City, Mexico), Mexican architect whose innovative works often merged geometry with nature. She was committed to collaboration as an essential feature of her work. Bilbao shared her interest in architecture and urban planning with a number of family members. One of her grandfathers was minister of urban development for the Spanish government before Francisco Franco came to power, and he moved his family to Mexico. Bilbao's own predilection for architecture was evident from her girlhood, when she took the gift of a Barbie doll as an excuse to build a city in which Barbie could live. (By her own account, she never played with the doll itself.) She later attended the Universidad Iberoamericana (UIA, also called Ibero), a Jesuit university in Mexico City, where she took a degree in architecture and urbanism (1996), having become passionate about the power of architecture to transform.

In 1998–99 Bilbao worked as an adviser for urban projects at the Ministry of Urban Development and Housing, Mexico City. Somewhat discouraged by the bureaucratic nature of that work, she struck out on her own, cofounding an architectural think tank called Laboratorio de la Ciudad de México (LCM) in 1999 and an urban research centre, MXDF, in 2004. In the latter year she also started her own firm, Tatiana Bilbao S.C. In 2005 she was appointed professor of design at her alma mater. Bilbao's first project was a collaboration with artist Gabriel Orozco on his beach house near Puerto Escondido. In the following years she honed a distinctly interdisciplinary approach, working with artists such as Ai Weiwei on projects in China and Spain, as well as in Mexico. In Sinaloa state, for example, she worked within the Culiacán Botanical Garden (designed by engineer and gardener Carlos Murillo) to install art works and necessary service buildings. She organized the gardens into an organic whole by connecting the natural and built environments with pathways inspired by the branches of a tree. Another of her projects for Culiacán was a multiuse biotechnology facility, Bioinnova (completed in 2012), that Bilbao and her team designed for the Monterrey Institute of Technology's campus there. Like Bilbao's other work, Bioinnova was inspired by a mix of geometry and nature—in this case, the form of a tree. Rather than producing a standard Modernist glass-and-steel block, Bilbao stacked four identical rectangular floors and shifted them in relation to one another to create overhangs and terraces; she also rotated one floor 90°, completely changing its orientation. The resulting staggered structure was functional as well as eye-catching, both inside and out. The concrete interior walls that formed the internal structure (or "tree trunk") of Bioinnova were painted a different sensuous tropical colour on each floor. The glass outer walls too were a different tint on each floor, and the mesh that covered the facade of the rotated floor further set it apart. The whole was pleasing and well-integrated, an inspiration for viewer and user alike.

At the Chicago Architecture Biennial 2015, Bilbao presented a "flexible building prototype," using modular components that could be adjusted to fit the number of inhabitants but could also be expanded by units to meet the needs of a growing family. Materials and layouts could also be varied to suit climate. The design was commissioned by the Mexican government to help allay the country's housing shortage with low-cost solutions. Bilbao also built several mixed-income residential buildings (2018) in La Confluence, a former industrial district in Lyon, France. Her designs for private homes responded to their environment in often astonishing ways, as in Los Terrenos (2016), Monterrey, Mexico, a dwelling with a mirrored glass envelope that reflects the wooded location. Bilbao was the recipient of various awards, and in 2010 she was named an Emerging Voice by the Architectural League of New York. In addition to the projects mentioned, Bilbao built a funeral home, a pavilion, a music hall and sports centre, and structures along a pilgrimage route.

### 1.7.6 Claude-Nicolas Ledoux (French architect)

Claude-Nicolas Ledoux, (born March 21, 1736, Dormans-sur-Marne, Fr.—died Nov. 19, 1806, Paris), French architect who developed an eclectic and visionary architecture linked with nascent pre-Revolutionary social ideals. Ledoux studied under J.-F. Blondel and L.-F. Trouard. His imaginative woodwork at a café brought him to the notice of society, and he soon became a fashionable architect. In the 1760s and early '70s he designed many private houses in an innovative Neoclassical style for the higher social circles in France. Among such few surviving works are the Pavilion Hocquart (1764–70), the Château de Bénouville, Normandy (1770), and the famous chateau for Madame du Barry at Louveciennes (1771–73).

In the mid-1770s Ledoux took on the planning for a new saltworks and its surrounding town at the Salines de Chaux, at Arc-et-Senans. He devised a radial concentric plan for the settlement, with rings of workers' dwellings enclosing a central salt-extraction factory. Less than half of the project was completed, but the remaining structures show Ledoux's striking simplifications of cubes and cylinders to create squat, massive, boldly rusticated (rough-hewn) versions of classical building types. His layout of the town to both facilitate economic production and ensure healthy and happy conditions for the workers anticipated similar planning efforts by Robert Owen and other 19th-century Utopian socialists.

Ledoux's Theatre of Besançon (1771–73) was a revolutionary design in its provision of seats for the ordinary public as well as for the upper classes. The private houses he designed in the 1780s had brilliantly eccentric features, including odd layouts, discontinuous elevations, and a striking use of Doric architectural elements. Ledoux's most important public project in the last phase of his career was to design 60 tollhouses situated at the city gates of Paris. He turned what might have been modest customs offices into a series of monumental gates and other structures called the Portes de Paris. Of the 50 such tollhouses, or barrières, actually built (1785–89) in the four years preceding the French Revolution, only four, including the famous Barrière de la Villette, still survive. In the barrières Ledoux took his interest in squat, colossal geometric forms to its furthest extent, fashioning rotundas, Greek temples, porticoes, and vaulted apses with rusticated masonry and Doric columns. The cost of these buildings proved ruinous to the public treasury, however, and he was dismissed from his project in 1789. Many of the barrières were subsequently torn down by mobs of resentful taxpayers during the Revolution. Ledoux himself was arrested during the Terror, and this event and the deaths of several members of his family ended his active career as an architect. After his release he spent his last years writing and compiling L'architecture considérée sous le rapport de l'art, des moeurs et de la législation (1804; "Architecture Considered with Respect to Art, Customs, and Legislation"), which contains his own engravings of his works. Ledoux was the most prolific, productive, and original architect of late 18th-century France. The powerful and brilliantly simplified geometry of his buildings held little appeal for the following generations, however, and wholesale demolitions and vandalism during the 19th century left only a handful of his works still standing. Among them is his salt works at Arc-et-Senans, which UNESCO designated a World Heritage site in 1982.

### 1.7.7 Herod (King of Judaea)

Herod, byname Herod the Great, Latin Herodes Magnus, (born 73 BCE—died March/ April, 4 BCE, Jericho, Judaea), Roman-appointed king of Judaea (37–4 BCE), who built many fortresses, aqueducts, theatres, and other public buildings and generally raised the prosperity of his land but who was the centre of political and family intrigues in his later years. The New Testament portrays him as a tyrant, into whose kingdom Jesus of Nazareth was born.

### Family And Early Life

Herod was born in southern Palestine. His father, Antipater, was an Edomite (a Semitic people, identified by some scholars as Arab, who converted to Judaism in the 2nd century BCE). Antipater was a man of great influence and wealth who increased both by marrying the daughter of a noble from Petra (in southwestern Jordan), at that time the capital of the rising Arab Nabataean kingdom. Thus, Herod was of Arab origin, although he was a practicing Jew. When Pompey (48–106 BCE) invaded Palestine in 63 BCE, Antipater supported his campaign and began a long association with Rome, from which both he and Herod were to benefit. Six years later Herod met Mark Antony, whose lifelong friend he was to remain. Julius Caesar also favoured the family; he appointed Antipater procurator of Judaea in 47 BCE and conferred on him Roman citizenship, an honour that descended to Herod and his children. Herod made his political debut in the same year, when his father appointed him governor of Galilee. Six years later Mark Antony made him tetrarch of Galilee.

## King of Palestine

In 40 BCE the Parthians invaded Palestine, civil war broke out, and Herod was forced to flee to Rome. The senate there nominated him king of Judaea and equipped him with an army to make good his claim. In the year 37 BCE, at the age of 36, Herod became the unchallenged ruler of Judaea, a position he was to maintain for 32 years. To further solidify his power, he divorced his first wife, Doris, sent her and his son away from court, and married Mariamne, a Hasmonean princess. Although the union was directed at ending his feud with the Hasmoneans, a priestly family of Jewish leaders, he was deeply in love with Mariamne. During the conflict between the two triumvirs Octavian and Antony, the heirs to Caesar's power, Herod supported his friend Antony. He continued to do so even when Antony's mistress, Cleopatra, the queen of Egypt, used her influence with Antony to gain much of Herod's best land. After Antony's final defeat at Actium in 31 BCE, he frankly confessed to the victorious Octavian which side he had taken. Octavian, who had met Herod in Rome, knew that he was the one man to rule Palestine as Rome wanted it ruled and confirmed him king. He also restored to Herod the land Cleopatra had taken.

## Construction of the Second Temple and role in the story of Jesus

Herod endowed his realm with massive fortresses and splendid cities, of which the two greatest were new, and largely pagan, foundations: the port of Caesarea Palaestinae on the coast between Joppa (Jaffa) and Haifa, which was afterward to become the capital of Roman Palestine; and Sebaste on the long-desolate site of ancient Samaria. At Herodium in the Judaean desert Herod built a great palace, which archaeologists in 2007 tentatively identified as the site of his tomb. In Jerusalem he built the fortress of Antonia, portions of which may still be seen beneath the convents on the Via Dolorosa, and a magnificent palace (of which part survives in the citadel). His most grandiose creation was the Temple, which he wholly rebuilt. The great outer court, 35 acres (14 hectares) in extent, is still visible as Al-Ḥaram al-Sharīf. He also embellished foreign cities—Beirut, Damascus, Antioch, Rhodes—and many towns. Herod patronized the Olympic Games, whose president he became. In his own kingdom he could not give full rein to his love of magnificence, for fear of offending the Pharisees, the leading faction of Judaism, with whom he was always in conflict because they regarded him as a foreigner. Herod undoubtedly saw himself not merely as the patron of grateful pagans but also as the protector of Jewry outside of Palestine, whose Gentile hosts he did all in his power to conciliate. Unfortunately, there was a dark and cruel streak in Herod's character that showed itself increasingly as he grew older. His mental instability, moreover, was fed by the intrigue and deception that went on within his own family. Despite his affection for Mariamne, he was prone to violent attacks of jealousy; his sister Salome (not to be confused with her great-niece, Herodias's daughter Salome) made good use of his natural suspicions and poisoned his mind against his wife in order to wreck the union. In the end Herod murdered Mariamne, her two sons, her brother, her grandfather, and her mother, a woman of the vilest stamp who had often aided his sister Salome's schemes. Besides Doris and Mariamne, Herod had eight other wives and had children by six of them. He had 14 children. In his last years Herod suffered from arteriosclerosis. He had to repress a revolt, became involved in a quarrel with his Nabataean neighbours, and finally lost the favour of Augustus. He was in great pain and in mental and physical disorder. He altered his will three times and finally disinherited and killed his firstborn, Antipater. The slaying, shortly before his death, of the infants of Bethlehem was wholly consistent with the disarray into which he had fallen. After an unsuccessful attempt at suicide, Herod died. His final testament provided that, subject to Augustus's sanction, his realm would be divided among his sons: Archelaus should be king of Judaea and Samaria, with Philip and Antipas sharing the remainder as tetrarchs.

### 1.7.8 Emmanuel Héré de Corny (French architect)

Emmanuel Héré de Corny, (born Oct. 12, 1705, Nancy, Fr.—died Feb. 2, 1763, Lunéville), French court architect to Stanisław Leszczyński, duke of Lorraine, best known for laying out the town centre of Nancy, a principal example of urban design in the 18th century. Little is known of Héré's training. Stanisław, the former king of Poland and father-in-law to Louis XV, was made duke of Lorraine in the 1730s. He commissioned Héré to unite the medieval and Renaissance sections of Nancy, separated at that time by a moat and the remnants of fortifications. Héré's resulting work, begun in 1752, consists of three areas: the Place Royale (now the Place Stanislas), the Place de la Carrière, and the Place du Gouvernement. These interconnected areas form a series of squares and promenades lined with buildings or ringed by colonnades. The squares are tastefully embellished with lines of trees and with sculptures and fountains. The layout created pleasant vistas, improved traffic patterns, and provided the basis for the future rational development of the city. In creating this masterful project in a provincial capital and on a limited budget, Héré established himself on the level of accomplishment of the most notable city planners of his time. Among his other works, almost all at Nancy, are the church of Notre-Dame-de-Bon-Secours (1738–41) and the Hostel of Royal Missions, erected for the Jesuits between 1741 and 1743.

# 2

# URBAN PLANNING AND URBAN DESIGN

## 2.1 URBAN PLANNING AND URBAN DESIGN
### AN INTRODUCTION

**What Is Urban?**

The term "urban" has been a serious limitation on the practice as it implies both that urban designers only consider intense urban environments orpursue the objective of urbanizing of everything they touch. Of course, this is far from the truth. Urban design has proven useful in rural areas (See Randall Arendt Rural by Design, 1994) and provided assistance achieving environmental resource management goals. (See Figure 5.) However, a suitable alternative nomenclature has been elusive. "Community design", favored by many, implies a certain scale and focus of activity. "Civic design" tends to mean large public infrastructure, and "environmental design" skews the term toward environmental protection. So, urban designers, especially those working insmaller communities have taken to explaining how the term is applied to their specific situation. For my purposes here, urban and the purview of an urban design approach includes all environments directly influenced by humans.

**What is Design?**

There are several dictionary entries for the verb "design".

    i.   To create, fashion, execute, or construct according to plan, 2. To conceive and plan out in the mind. (Merriam Webster On-Line Dictionary)

    ii.   To make or draw plans for something (Cambridge Dictionary).

    iii.   To plan and fashion artistically or skillfully,

    iv.   To form or conceive in the mind. (Webster's Unabridged Dictionary of the English Language).

Implied in these definitions are four aspects of the design process that characterize urban designer's work and make it a critical component of the planning and design community.

Since the existence of humankind, planning was among the main issues to deal with; this is so because planning helps him to organize his activities and to predict his future. It is on the basis of this that cities have also been put into account in planning. However, as the human society is growing up, this development is bringing some problems to our earth since there is over exploitation of the world resources. Due to this over exploitation of the earth, natural resources; and other mal- human activities (socio-economic), such as pollution and waste problems, lose of natures conservation, biodiversity, ecology and green coupled with the emergence and dangers of climate change, global warming and its negative implications on living quality in recent times and in future to the whole world, there is a growing interest in sustainability, sustainable development and its incorporation into plans of all sorts. Despite this issues of global warming, researches have shown that green open space can be one of the solutions of that problem. Green open space has effect on microclimates. Trees and planting can result in the reduction of peak summer temperatures by up to 5° Celsius. Thus, trees can be included in the City's street scene designs wherever possible, to provide shade and cooling and consequently reduce global warming. Basing on the above said, we found important to discuss about urban design and urban planning: Green Open Space.

## 2.2 CONNECTION BETWEEN URBAN PLANNING AND URBAN DESIGN

According to Peter Hall, 2002, the verb "to plan" and the noun "planning" and planner, have in fact only the second general group of meanings: they do refer to the art of drawing up a physical plan or design on paper. They can mean either "either "to arrange the parts of" or "to realize the achievement of "or" more vaguely, to "intend". The most common meaning of planning involves both the first two of these elements; planning is concerned with deliberately achieving some objectives and it proceeds by assembling actions into some orderly sequence. It is on the basis of this that Gallion and Eisner, 1993, defined urban planning as an attempt to create an orderly development in urban areas and reduce social conflicts and economic conditions that would endanger the lives and property.

On the other hand, urban design is about how to recapture certain of the qualities (qualities which we experience as well as those we see) that we associate with the traditional city: a sense of order, place, and continuity, richness of experience, completeness and belonging. Urban design lies somewhere between the broad-brush abstractions of planning and the concrete specifics of architecture. As we can see it through the definitions, there are some similarities and differences between these two subjects. urban planning has a role to make sure that a city is working and functioning well that is why urban and regional planning is focusing on various issues such as economical, social and environmental issues while, urban design is more focusing on aesthetical values like sense of place, building character, pedestrian design and design of public space. Furthermore, sometimes urban planners can become very micro like urban designers. They can make some designs of the cities which is the main task of urban designers. In revenge urban designers can make plans for policies and guidelines. Because urban planning and urban design are very wide, we found important to discuss about green open space as a focus point.

## 2.3 GREEN OPEN SPACE

Green Open Space is the region or ground surface area which is dominated by plants that are promoted to a particular habitat protection functions, and / or facilities neighborhood / city, and / or network security infrastructure, and / or agricultural cultivation. Nowadays, the concept" open space" in complex matrix is not limited only to the urban park and preserves but also non park-non natural-places. Public spaces such as streets, school yards, outdoor sport complexes, cemeteries, and public squares are important green open spaces.

### 2.3.1 Why Plan And Design For Green Open Space?

The process of getting everyone together to think about community needs is worthwhile endeavor in itself. An urban open space plan is much more than a land acquisition plan. It can make a wide variety of recommendations about the future of a country. So, we have to plan for a green open space because:

- ⊙ A green open space plan is the flip side of a development plan. After identifying important green open spaces, it will be much more apparent where development should occur.

- ⊙ It recommends land use regulations that will help to protect the community from uneconomic and inefficient sprawl.

It is for this reason that it is essential for urban planners to determine the function of green open space in order to increase its value (such as water conservation, wetland area, city lungs). When you talk about why plan and design for green open space, it is essential to look at the size and levels of green open space.

### 2.3.2 Green Open Space (Size And Level)

It exists two types of green open space: rural open space and urban open space. Rural open space is made of habitat, recreation, health/ safety (flooding/seismic), agriculture/ rangelands, river and stream parkways while urban open space is constituted by recreation, trails and parkways, stream and canal corridors, natural resources and public space. However, it is too difficult to determine an international size of green open space by different levels because every country has its policies, own physical characteristics and culture but we have an example from United Kingdom which can be a best practice.

- ◉ The United Kingdom Accessible Natural Green Space Standard (ANGS) mentioned that:

- ◉ No person should live more than 300 m from their nearest area of natural green space of at least two (2) hectares in size;

- ◉ There is provision of at least two (2) hectares of Local Nature Reserve per 1,000 population;

- ◉ That there should be at least one accessible 20 ha site within two (2) km from home

- ◉ That there should be one accessible 100 ha site within five (5) km;

- ◉ That there should be one accessible 500 ha site within (10) km.

We can not only discuss about the size and levels of green open space in urban planning and urban design, we need also to recognize the functions of green open space because the latter are always planned for certain purposes.

### 2.3.3 Function of Green Open Space

Green open spaces are vital part of landscapes with its own specific set of function. Open spaces (natural or manmade) contribute to the quality of life in many ways (Burke and Ewan, 1999). Beside important environmental benefits (such as improvement of the quality of air, soil and water, decrease of noise levels, reduction of thermal amplitude variations, protection against the winds, waste Management, improvement of the infiltration and drainage of storm water, reduction of flood risks), these areas provide social psychological services (such as Recreation and Leisure, Increasing physical and Psychological well-being, Sociability) which are critical for the livability of the city and well being of urbanites. Green open space as places to celebrate cultural diversity, to engage with natural processes and to conserve memories. Green open space has also economic function: it promotes the image of the city, increase the selling point. It contributes ecologically because it diminishes the process of erosion and promotes biodiversity. These above functions can be combined each other. For example in Houten, we saw that green open space is combined with wet land area. The functions of green open space are water conservation and recreation. Also, green areas are used to encourage people to cycling. This is done by planning green open space along the bicycle pathways. Here, green space has multifunctional purpose like encouraging cycling and enhancing community health. As it is stated above, today, green open space is mainly planned with a purpose of fighting against global warming; this is why we found important to talk about its role in combating against this worldwide issue.

## 2.4 THE ROLE OF GREEN OPEN SPACE IN FIGHTING AGAINST GLOBAL WARMING AND CLIMATE CHANGE

There is a growing consensus that global warming is one of the greatest threats facing humanity. Different researches have shown that greenhouse gases are the first to keep the earth warm, human use of fossil fuels is the main source of excess greenhouse gases.

By driving cars, using electricity from coal-fired power plants, or heating our homes with oil or natural gas, we release carbon dioxide and other heat-trapping gases into the atmosphere. Deforestation is another significant source of greenhouse gases, because fewer trees mean less carbon dioxide conversion of oxygen. This is why some scientists say that: "The bigger are the cities, the more the urban citizen is deprived of contact with the environment, the more he despises or simply ignores the other elements of ecosystem. He becomes more and more hostile and the men look for refuge during more time inside artificial environment". This author wanted to express that the growing of the city affects the environment because the forest, trees or green in general are replaced by urban infrastructures like houses, streets, public building, etc. This idea is also supported by Sing Chew, 2001 in his book: world ecological degradation. Accumulation, urbanization and deforestation, he said that: "In all the process of urbanization, depending on the nature and levels of consumption and production, generates ecological degradation when the nature becomes extremely exploitative".

## 2.5 THE URGENCY OF URBAN PLANNING TODAY

Within a few decades' time, we can expect the planet to become more crowded, resources more precious, and innovative urban planners increasingly important. By midcentury, the global population will likely top nine billion, and more than half will live in cities. What will these cities look like? Will we have the resources to power them and comfortably provide for their residents? Will global urbanization harmonize with efforts to curb climate change and secure a sustainable future, or are these forces hurtling towards a head-on collision? The TED speakers featured in Ecofying Cities underscore the urgency, but also suggest that some optimism's in order as they outline the issues and offer imaginative solutions. There's no single reason for or response to the complex environmental, economic and social challenges that are part of our future in cities. They call for multiple approaches, originating from different sources — individuals, communities, governments, businesses — and deployed at different levels — in the home, the neighborhood, the city, region, nation and across the globe — to respond to the challenges at hand. As Alex Steffen reminds the urban planners, architects, designers, elected leaders and others involved in the effort, "All those cities are opportunities."

## 2.6 URBANISM AND THE ENVIRONMENT

For centuries, successful city-building has required careful attention to the environmental consequences of urban development. Without this, as Jared Diamond demonstrated in *Collapse: How Societies Choose to Fail or Succeed*, a city inevitably ended up fouling its nest, thus entering a spiral of epidemics, economic hardship, decline and, ultimately, oblivion. Civilizations evolved different ways of dealing with environmental considerations — some with more success than others. For example, thanks to elaborate aqueducts and sewer systems, the Romans were able to build and sustain for centuries large cities that featured a reliable public water supply and state-of-the-art public health conditions.

In other civilizations, however, residents simply abandoned cities when they could no longer rely on their environment to supply the resources they needed. Often this was a direct result of their own activities: for example, deforestation and the attendant erosion of fertile soil, epidemics due to contaminated water and, with the advent of coal-fired industrialization, air pollution. Urban planning got its start as a profession largely dedicated to averting different types of crises arising from urban growth and providing conditions for public health. This was particularly true in the many 19th century European and North American cities transformed by industrialization and unprecedented rates of population growth. Rapidly deteriorating air and water quality made it necessary to introduce regulations to protect the health of the residents of these cities.

The planners' first-generation improvements included sewers, water treatment and distribution, and improved air quality through building codes and increased urban green space. It's especially remarkable today to think that these interventions were adopted in response to observable health consequences, but *without* knowledge of the contamination mechanisms at work: germ theory didn›t arrive on the scene until Louis Pasteur published his work in the 1860s. From the late 19th century onward Pasteur›s findings bolstered the case for even more urban sanitation improvements, particularly those designed to improve water quality. Starting in the 1950s, however, planners no longer narrowly targeted immediate health effects on urban residents as their chief environmental concern. Their work also absorbed and reflected Western society's deeper understanding of, and respect for, natural processes and growing awareness of the long-term environmental impacts of cities from the local to the planetary scale. Rachel Carson is often credited as the first to popularize environmentalism. Published in 1962, her landmark book *Silent Spring* sounded a warning call about how pesticides endanger birds and entire ecological systems. Soon after, air pollution became a rallying point for environmentalists, as did the loss of large tracks of rural and natural land to accelerated, sprawling development. Today, sustainable development and smart growth, which largely overlap and address multiple environmental considerations, enjoy wide currency; most urban planning is now based on these principles. Today, as we reckon with population growth, advancing rates of urbanization, and widespread recognition of climate change, we know that the cities of the future share a common destiny. The choices we make about how we build inhabit and maintain these cities will have global and long-term effects.

## 2.7 SUSTAINABLE DEVELOPMENT

In modern urban planning, there are two general categories of sustainable development. The first doesn't challenge the present dynamics of the city, allowing them to remain largely low-density and automobile-oriented, but still makes them the object of measures aimed to reduce their environmental load (for example, green construction practices). Ian McHarg spearheaded this approach as a way to develop urban areas in harmony with natural systems; the planning principles he formulated gave special care to the preservation of water and green space. His lasting influence is visible in many of the more enlightened suburban developments of recent decades which respect the integrity of natural systems. Today, the Landscape Urbanism movement promotes these same ideas. A second school of urban development focuses on increasing urban density and reducing reliance on the automobile. This approach advocates transit-oriented and mixed-use development along pedestrian-friendly "complete streets." On a regional scale, it aims to reduce sprawl by creating a network of higher-density multifunctional centers interconnected by public transit. Today, it's common for plans with a metropolitan scope to follow this approach.

## 2.8 STUDYING THE CITY

Cities are arguably the most complex human creation (with the possible exception of language) so it's not surprising that we study them at multiple scales and from diverse perspectives. We can approach cities through a narrow focus on an individual building or a neighborhood, expand the investigation to consider a metropolitan region in its entirety, or study the global system of cities and its interconnections. What's more, we can think about cities as built environments, social networks, modified ecologies, economic systems and political entities. Aware of the multiple ways that we engage with cities, the Romans had two words to refer to them: *urbs* referred to the physical city with its wall and buildings, and *civitas*, the city as a collection of residents.

Ecofying Cities explores urban areas at different scales. In some cases, the TED speaker focuses on a neighborhood project, like The High Line in Manhattan; others describe city-wide transformation, as in Curitiba, Brazil, or a regional or national initiative like China's plan for a network of eco-cities to house its growing urban population. Likewise, the talks explore cities from different disciplinary perspectives including urban planning, urban design, transportation planning, architecture, community organization and environmental science. What unites them all? A commitment to sustainability and a belief that sustainability is more about creating positive effects rather than reducing negative impacts. The message emanating from Ecofying Cities is one of complexity, optimism and uncertainty. We can't be sure that the changes these speakers suggest will be enough to help us balance supply and demand in the sustainability equation. But we can expect that their ideas and efforts will improve the built environment — as well as quality of life — in cities, thereby providing hopeful perspectives for a sustainable future. Let´s begin with writer and futurist Alex Steffen´s TEDTalk "The Sharable Future of Cities" for a look at the interplay between increasing urban density and energy consumption.

## 2.9 URBAN DESIGN: FROM DEFINITION TO APPROACH

As articulated by Professor Meyer Wolfe, Urban design is the manipulation of the physical environment, in a way that

- ◉ Pursues multiple objectives,

- ◉ For multiple clients (including affected membersof the public), that

- ◉ Addresses the way people perceive and behave in their surroundings,

- ◉ Considers the implications of form-giving actions (including the environmental and ecological consequences) at a range of scales (sometimes from the individual to the regional), and

- ◉ Is conducted through an explicit public decision-making process that:

- ◉ Offers the pubic the opportunity to participate in the process in a meaningful way,

- ◉ Identifies goals and objectives,

- ◉ Analyzes existing conditions,

- ◉ Explores alternate concepts and solutions,

- ◉ Evaluates those options with respect to project goals and public values,

- ◉ Selects the preferred alternative or combines preferred elements into a synthesized concept, and

- ◉ Includes an implementation strategy.

This is a powerful definition for three reasons. First, it carries a set of implicit values that all applicable urban design activities should pursue. That is, to be signified as urban design, an activity must be conducted in an inclusive public process that addresses the multiple objectives of those who are affected. This balancing of various interests should lead directly to the pursuit of fair and equitable solutions. By emphasizing effects on human perception and behavior, practitioners will hopefully avoid some of the inhuman and dysfunctional spaces that have been created in the name of city planning. And, considering a proposed action's impacts at a range of scales, will help urban designers connect their efforts to larger (and sometimes more intimate) physical and environmental implications as well as broader policy objectives.

Secondly, the definition proposed above provides a useful checklist for designers, planners, engineers and other practitioners to use so that they are addressing urban design's inherent values noted above. While it may seem superfluous to pursue "urban design" for its own sake, tracking activities relative to the definition'skey elements will help to keep the project focused on relevant policy goals and physical objectives.

Third, by describing a rational participatory process through which to pursue the discipline, the definition provides a clear methodology for applying urban design concepts. This aspect will be discussed in the section describing an explicit public decision making process.

## 2.10 COMMON URBAN DESIGN ACTIVITIES

Figure. 1 below illustrates some of the activities that are traditionally considered under the purview of urban design. Most of them are conducted by teams of professionals, usually led by an architect, landscape, architect or planner with urban design expertise.

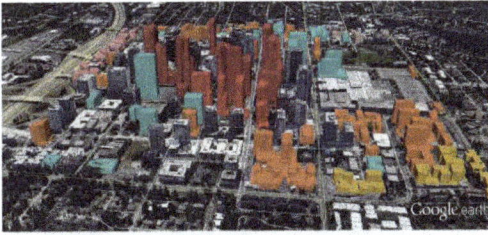
Metropolitan center urban design plans

Vision: A Linear Community

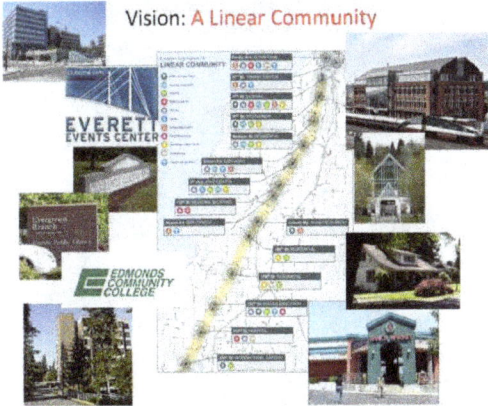
Urban corridor plans

Waterfront facilities

Rural towns and corridors

**Fig. 1: a, Some typical urban design projects.**

Towards An Approach: As above attempts to cordon off urban design as a discreet discipline, it must be acknowledged that such a bright line definition is very porous – that many of its defining activities and principles are common with other city building, planning and design activities - and that many of these same activities are integrated into other planning efforts addressing a broader range of public goals. This is one reason many definitions of urban design focus on the physical aspects of urban form as the most important definitional identifier.

However, a distinct urban design approach which includes the elements and process described above is still needed within the spectrum of urban planning and design fields for the reasons noted above. Namely, an urban design approach carries with it fundamental principles and responses to human behaviors, provides a process and conceptual framework for implementing those principles, and allows more expansive use of the tools that urban design brings to a physical planning challenge.Viewing urban design as an approach to meeting physical planning challenges rather than as a discreet discipline further blurs he lines defining the term. However, as will be seen, it greatly expands its use in city and community building and adds tools that the broader spectrum of planners, architects and landscape architects can use.

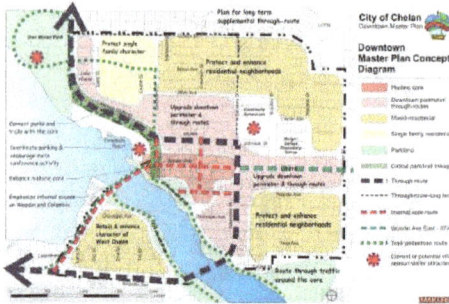
Downtown and neighborhood Plans

Waterfront plans and redevelopments

Entry weather protection

Window (bottom of window generally above pedestrian eye level)

Outdoor living space

Threshold element defining private property and public realm

6' min
3' min

6' min (recommended)
10' (Generally necessary for public/private transition)

Design guidelines and form-based regulations

Transit oriented development

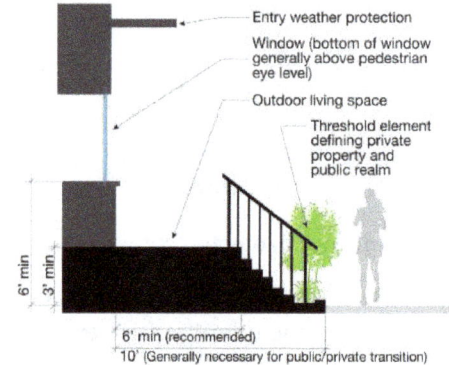
Parks and open spaces

Elements of large infrastructure projects

Streets and pathways

Transit facilities

**Fig. 1: b, Some typical urban design projects.**

## 2.11 ELEMENTS OF URBAN DESIGN: EMERGING CHALLENGES AND NEW IDEAS

This section considers the definition's elements in more detail and explores their implications for today' practice. There are several sections that digress from the primary topics, but, hopefully they amplify the salient points and provide useful insights related to professional urban design and planning practices. Many of the examples are culled from my experience as a professional architect and urban designer.

### 2.11.1 Manipulating the Physical Environment

First, it should be clear that "manipulating the physical environment", encompasses a broad set of activities in a wide range of physical settings. "Manipulation" may include direct physical design and construction, regulatory measures to guide physical changes over time, economic and community development efforts, regional growth strategies, and institutional measures such as funding programs that ultimately result in physical actions. The Urban Waterfront Policy Analysis example in Figure 2 demonstrates that urban design techniques can be used to address broad policy objectives even if there is no direct physical action.

Similarly, the term "physical environment" is to be broadly interpreted and include intense urban settings, but also local communities, suburban centers, small towns, and rural areas. And, urban design tools have also been very useful in addressing ecological planning and restoration in both urban and wilderness settings.

### 2.11.2 Pursue Multiple Objectives

Given the immediacy of numerous challenges currently facing cities – transportation gridlock, housing shortages, homelessness, gentrification, etc. – it is not surprising that planners are directed to focus on single purpose projects targeted to specific needs. In many ways, such responses are appropriate and necessary. However, an urban design perspective that incorporates multiple objectives and form-based measures can facilitate achieving the original objective by leading to a more comprehensive approach.

For example, the City of Seattle has recently adopted an ambitious mandatory inclusionary zoning process that increases the development capacity of private lands in exchange for affordable housing requirements. Under the program, property owners will benefit froman "up-zone" that substantially increases their development capacity in exchange for providing affordable housing as part of the development or a fee in lieu which the City can use to develop affordable housing off-site. Several community groups raised issues related to the impacts of larger development to neighbors who now face the loss of privacy, sunlight and views due to the increased bulk of new development – not to mention the need for more open space and public infrastructure to accommodate the influx of new residents. Although this program was established under the title, Housing and Livability Agenda, the City decided not to address the livability aspects of the program through any measures such as design guidelines, streetscape improvements, or increased multi-modal transportation or open space needs. This was a lost opportunity, but the program successfully proceeded to adoption, and hopefully such measures will be added over time to address the increased intensity of development.

Similarly, urban designers should assertively champion, for example, a consideration of community building and urban form objectives during transit station planning, or street typologies that account for adjacent land uses and building forms; issues that are not considered in transportation specific street classifications or even "complete street" models.

### 2.11.3 Address the Objectives of Multiple Clients

One way to illustrate the importance of balancing the objectives of all people affected by an urban design action is to contrast the urban design approach with large site master planning, which is typically initiated by a single entity for its specific and individual purposes. It may be that the property owners master planning their properties engage the public, but in the end, they will typically gravitate to their own interests – a perfectly valid thing to do, but it is not urban design as defined here.

This difference is illustrated by the Virginia Mason Medical Center (VMMC) Institutional Master Plan which is required and must be approved by the City of Seattle before the before any major changes to the medical center's campus can be initiated. At first the VMMC master planners focused on the complex's internal workings with scant regard for community interests. When community members voiced strong opposition to this approach and threatened to derail the Institutional Master Plan process, the VMMC added an urban design team whose task was to faithfully engage the public and address their concerns. The team negotiated with community members to mitigate impacts due to proposed construction and improve the public realm in and around the campus. After several rounds of exploring trade-offs and design solutions (in the reflection-in action-mode noted above), the community endorsed the Institutional Master Plan proposal so that VMMC could begin implementation.

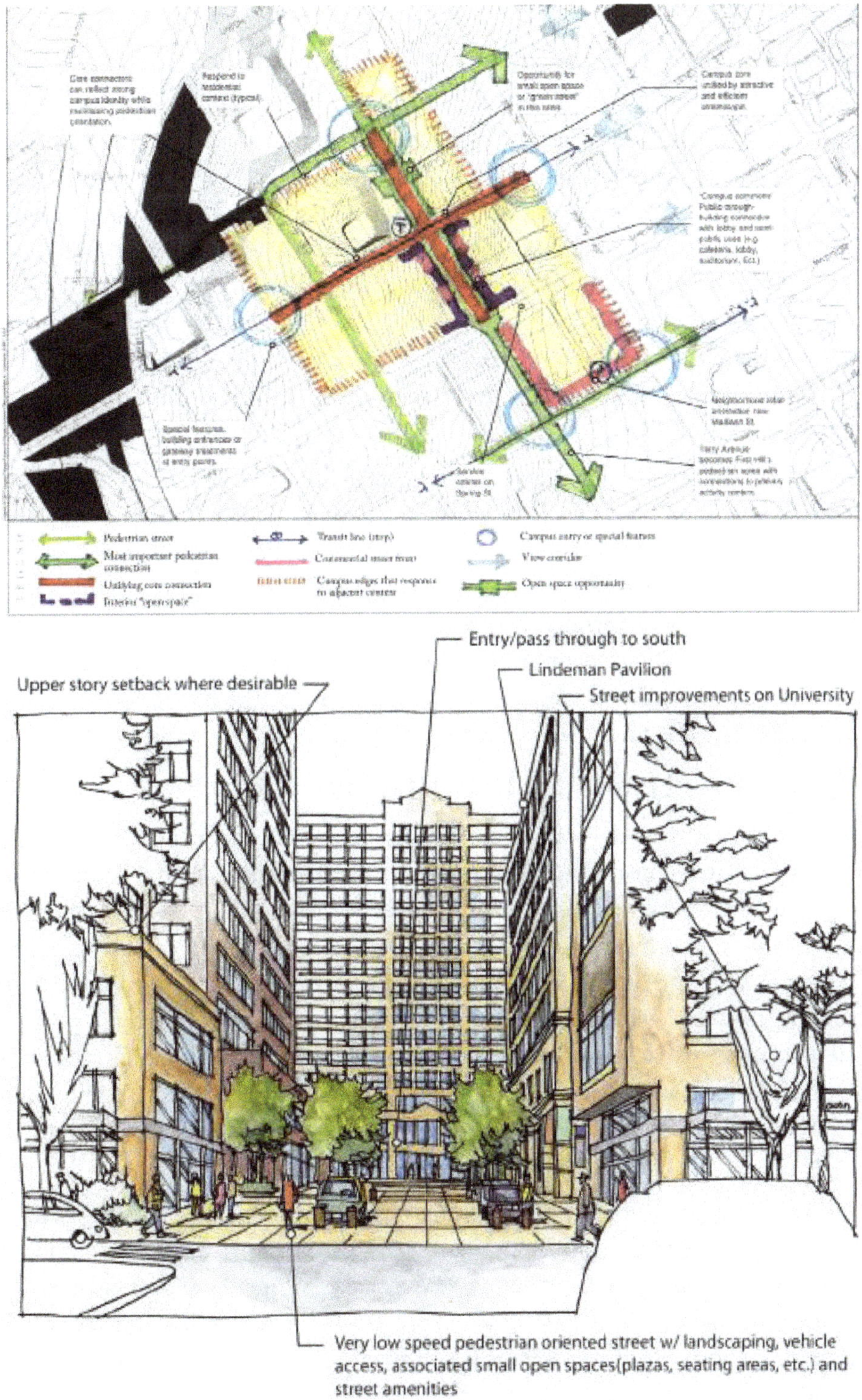

Fig. 2: A few of the design measures added to the VMMC Institutional Master Plan.
Top: an early diagram of community improvement elements. Bottom: One of several
renderings illustrating the community's requirements.

## 2.12 A CHALLENGE: FAIR AND INCLUSIVE PUBLIC ENGAGEMENT

Of course, public engagement of this sort has almost become the norm in community planning activities. Professionals have developed numerous techniques to elicit comment and work with groups toward common goals orto successfully resolve conflicts. But, too often, especially in diverse communities, these public input efforts do not adequately engage underrepresented populations who are unable or reluctant to participate in public meetings. Too often, the recommendations to public officials that emerge out of public engagement processes are shaped by those with the capabilities to make themselves heard while others, generally those with fewer resources, cannot or choose not to participate. This can be a severe limitation on the profession's ability to provide for fair public policies and actions.

In the past, many designers and planners working with the public assumed that if people were interested in a specific issue, they would participate. We know now that this is not necessarily true and that new outreach methods must be employed. Many people experience barriers to actively voicing their interests, including: lack of time, language difficulties, distrust of government, lack of transportation, and cultural disposition. If we want equitable solutions, we must provide equitably accessible participation, and that means finding new ways to engage a broader spectrum of the public. While there is no magic bullet to address this challenge the avenues below seem promising.

## 2.13 NEIGHBORHOOD ORGANIZATIONS

Cities are finding that official staff-initiated outreach activities for individual projects are both costly and time consuming; and that they are often unsuccessful. An alternate and previously used approach, more common in Seattle during the 1970s through the 1990s was to assist local community organizations (councils, community improvement clubs, etc.) and enlist them to bean active part of outreach efforts. Local groups, such as PTAs, churches, schools, etc. with their social and organizational connections can get the word out and encourage participation more effectively than city staff, although staff support is also critical. The key is to ensure that the community groups are truly representative of their communities. If they could so demonstrate, the groups received City support and resources.

This place-based approach has garnered and lost favor from time to time, but at present, it appears that proximity and the social connections within a community do matter. That means that urban design, which addresses social interactions in the public realm, is important in the larger public engagement effort. While the value of local organizations' efforts has been demonstrated by the results of physical improvements – a new park here, an improved development proposal there– a real benefit is that these organizations can serve as an on-going conduit for communication and collaboration between individual community members and public officials. These connections are especially valuable when a local issue or concern arises.

### 2.13.1 Social Indicators

A number of municipalities and local governments across the nation are using GIS based information systems to analyze social indicators related to human health and safety, economic stability, education attainment, non-English speakers and quality of life factors And, the City of Seattle has recently developed an analysis identifying neighborhoods most vulnerable to displacement. Such analyses can identify trends and geographic hot spots that require special attention regarding community outreach.

### 2.13.2 Coordination with Social Service Providers

Too often, the role of organizations providing social and health services, such as human and mental health, children and youth programs, housing, etc. is not included as part of urban design projects, probably because urban design is so focused on physical design. However, there are quite a number of overlapping objectives, and greater collaboration should be initiated. This is especially true when attempting to outreach to and address the needs of under-represented groups. Social service providers can more easily identify those populations where their clients come from and the best way to engage them in the public process.

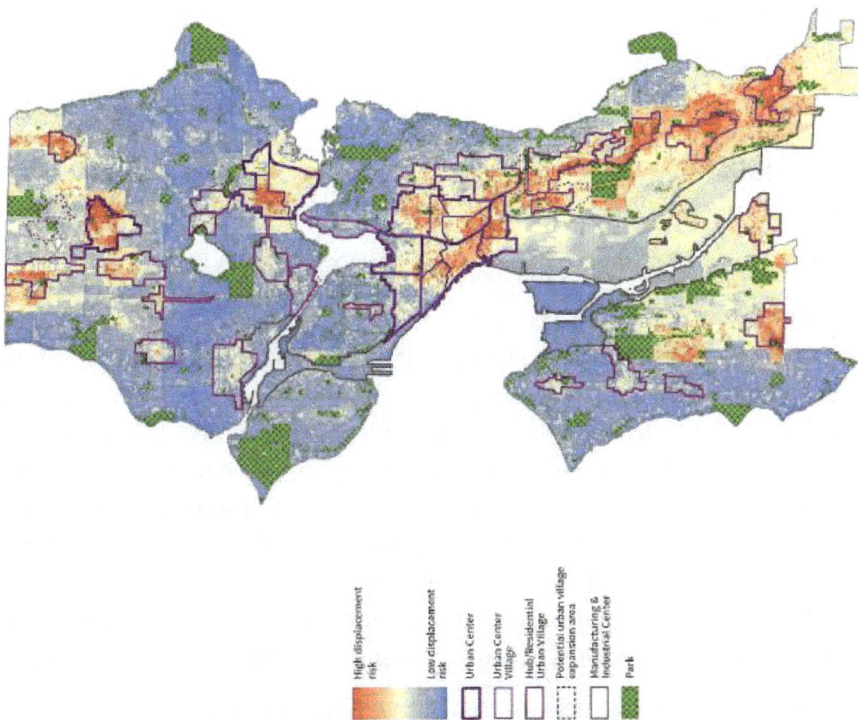

**Figure 3. Map from Seattle's Equitable Development Implementation Plan, 2016, illustrating relative vulnerability to displacement.**

## 2.14 ADDRESS THE SENSORY ENVIRONMENT

In shaping the human environment, whether that includes the design of streets, compositions of buildings, pastoral parks, neighbourhood centres or small towns, urban designers must understand and address how people and collections of people perceive and act within their environments. This is a core urban design capability not often addressed by other disciplines. Urban design research has produced very useful insights when considering person's immediate surroundings. However, there are other, broader aspects of human behavior, especially collective human behavior at a variety of scales that should be considered. More specifically, urban designers and those initiating research on this topic should considerthe following aspects of the human-environment relationship.

**Perception: Human responses to the sensory environment**

This topic addresses questions such as:

- ◉ What makes a street seem to narrow? What makes it perceived as a cohesive and comfortable space?

- ◉ How do we perceive spaces through which we travel? How can we design a roadway or pathway that offers a pleasant sequence of stimuli?

- ◉ Where do our eyes fix in and intense urban setting? Are the upper stories of high rises in our perceptions? Or, do we focus our attention at eye level, mostly noting the streetscape and the first few floors of buildings on near-by blocks? And, what does that say about how we write design guidelines for urban settings?

A related aspect of human perception that deserves special mention is the way we perceive linear spaces as we move through them; in other words, our sequential experiences. This has special relevance when we design streetscapes, highways or pathways and is a good example of translating perceptual preferences into urban design parameters.

## 2.15 CULTURAL BEHAVIORS AND PREFERENCES

For the most part, urban designers have tried to be sensitive to local populations byaccommodating or accentuating ethnic cultural activities and expressions through design. For example, designers have designed parks to accommodate ethnic festivals and the ways people from different cultures use space. This response to cultural behavior, associations and preferences might be broadened to consider culture in the larger, more inclusive sense. If culture is "the customs, arts, social institutions, and achievements of a particular nation, people, or other social group", is the Silicon Valley "culture" a real thing? As America divides itself along ideological and political lines, are we also creating different cultures? How should urban designers respond, if at all? The issues surrounding gentrification certainly have to do with rapidly changing cultures in a given neighborhood.

## 2.16 ECONOMIC AND FUNCTIONAL BEHAVIORS

On many urban design projects such as community plans, site development, transit oriented development planning, and other efforts where private development is an objective, economic behavior assumes importance because the market demand for various uses and the feasibility of site development will affect thetype, size, location and configuration of what gets built. For the types of projects identified above, it is generally necessary to consult a real estate economist or development specialist to:

- ⦿ Identify the development opportunities for new or re-development based on a market analysis,

- ⦿ Evaluate the feasibility of different development types and sites by conducting pro forma analyses of scenarios created by the urban designer or architect, and

- ⦿ Advise on regulatory measures, capital improvements and development incentives that would encourage the desired development.

## 2.17 BEHAVIORS RELATED TO LOCATIONAL PREFERENCES AND ACCESSIBILITY

Admittedly, researching and addressing the regional and sub-regional urban growth and circulation patterns seems a little far afield from the initial directive to link human behavior to environmental setting. However, these larger urban patterns are the result of human behaviors, preferences and movement. As demonstrated regional scale efforts are an important part of the design professions' purview, and the emerging geodesign tools are making his ideas even more powerful. And, given the rapid growth of metropolitan regions world- wide, understanding these patterns is of paramount importance if the planning and design disciplines are to be able to influence them. Addressing this topic does mean that urban designers must think in larger terms than what has traditionally considered urban design.

There are a number of design related concerns raised by the study of urban growth patterns For example, the VISION 2040 Regional Growth Strategy adopted by the Puget Sound Regional Council (PSRC), emphasizes a polynucleated constellation of urban centers linked by multi-modal transportation corridors. However, in actuality, central Seattle has encouraged and absorbed the preponderance of employment growth, exacerbating transportation problems and housing shortages. So, the question is, what tools can the planning and design professions bring to bear in managing intense urbanization in a more effective and equitable way. (See also the section on gentrification, displacement and equity, below.)

Other questions arise from transportation and access. As noticed from looking at a broad spectrum of cities, a city's geographic extent is limited by the time a person spends each day commuting to and from work. He found that, in general, the distance from a city's outer limits to its core has traditional been limited to the distance a commuter or shopper can travel from residence to their work place or central market in 30 minutes. When the primary transportation mode was by walking, As found most cities were less than about 3 miles across. With the advent of transportation by horse, cities grew in geographic size, and with automobiles and high capacity transit, cities typically can reach 30 to 50 miles across; given a 30 to 50 mph average speed. Obviously, transportation and communication technologies will continue to drive human behaviors. Urban designers in particular must identify those physical development models that will help communities adapt to macro-scaled changes and create livable and equitable conditions. One other aspect of transportation related behaviors is the time it takes for a person to access a particular discretionary destination, such as an entertainment venue, a friend's house ora park. Experts in the tourism industry say that as a general rule of thumb, people will make a one-way trip to an attraction or destination that that takes about one-fourth the time spent at that destination. Assuming that the rule of thumb applies for short trips – and this assumption should be verified as it is an important for planning in general – a person typically would-be willing to travel ½ hour for a 2-hour movie or dinner at a special restaurant. But what if congestion and parking difficulties increase that same trip to 45 minutes or an hour? Does that mean a reduction in business for the movie- theater, opera house, or specialty restaurant?

Such questions are significant for the structure of urban communities as well as the quality of life for their residents. They are questions for land use and transportation planners in general, but they also impact urban form at a variety of scales.

Social scientists, mathematicians, and urbanologists, as well as planners and designers are studying urban growth from a variety of perspectives. For example, Geoffrey West (2017) notes that from a mathematical point of view which examines the size and function based on both the theoretical "power law" and observed data, as cities grow larger they gain both an efficiency in per person infrastructure costs and economic and intellectual productivity. According to Miller, with every doubling of population, there is an effective 15% reduction in the need for new infrastructure and a corresponding 15% increase in human productivity, wealth and innovation.

That is, every time the population rises 100%, infrastructure requirements rise only 85% but productivity rises 115%. Unfortunately, there is also a 15% increase per person in crime, pollution and disease. And these relationships hold true for the great majority of cities within a given national economic and regulatory framework.

This means that there is an increased functional efficiency for larger urban centers, so one should not be surprised that population, resources and investment flow to our largest metropolises. This raises several questions, among them:

- Should society embrace this growth pattern? (The urban design question is. "What are the impacts of different macro-growth patterns on community livability?")

- How can rapid urban growth be managed more effectively?

- What urban forms are most effective in leveraging the efficiencies and minimizing the negative impacts of urban growth at this larger scale?

◉ Are some urban morphological forms more resilient than others? Do some forms encourage social contact more effectively?

Another avenue of research involves mathematical analysis based on an emerging understanding of complex adaptive systems to explore these questions. Using complexity science tools, mathematicians and scientists are measuring physical parameters of different urban patterns such as network connectivity, fractal structure, visual variety, information density and land use diversity. They then relate those metrics to urban design objectives such as resilience, accessibility, and livability. It appears that this work will be an especially exciting field of study for those willing to tackle the math – or team up with a friendly mathematician.

## 2.18 GENTRIFICATION, DISPLACEMENT AND SOCIAL EQUITY

Gentrification and displacement are particularly thorny issues arising out of the economic and land use development patterns discussed above. Urban design projects such as street, park, and community facility improvements have traditionally been used to enhance the livability and economic activity of poorer neighborhoods. In periods of low to moderate growth such strategies have often worked well. Local residents and businesses have been able to take advantage of increased investment and realized its benefits. During periods of rapid growth, however, a project such as a main street improvement or a rapid transit station can attract large scale outside investment resulting in whole sale, up-scale redevelopment that changes the neighborhood character and may displace residents and businesses who cannot afford the increased rents and prices.

Acknowledging that economic, racial and social equity issues must be addressed at a more fundamental, societal level, urban designers must be more sensitive to the social and economic impacts of their projects, particularly in poorer, under invested neighborhoods where public resources should otherwise be targeted. This will likely mean coordinating urban design improvements with a comprehensive suite of housing affordability, small business assistance and other measures to address the impacts of rapid redevelopment. Cities across the country are struggling to solve this issue, and it requires urban designers to remain sensitive to in their work. In conclusion there are many current urban design questions and challenges arising from the human-environment relationship. It would be helpful if existing knowledge about the way people perceive and use public spaces, identify gaps in our understanding and direct research to those missing pieces. A better understanding of the way people perceive their surroundings would help urban designers design more attractive, functional and safer streets – and craft more sensitive design guidelines for new development. Urban designers and related disciplines should work closely with economists and transportation planners to incorporate the latest information from those disciplines, and the study of geo-spatial behavior is becoming more crucial as cities adapt to rapid growth patterns transportation constraints, the internet and other emerging transformative trends.

## 2.19 CONSIDER THE IMPLICATIONS OF FORM-GIVING ACTIONS AT A RANGE OF SCALES

In simplest term of, this element directs the urban designer to look beyond the project boundaries to identify both the positive opportunities as well as the positive and negative impacts of the design intervention. The Delridge Triangle design presents a straight-forward example of this notion. The initial project entailed the redesign of a triangular piece of land consisting of the remaining right-of-way left over from a "Y" shaped intersection. The planning team, consisting of community members, architects, planners and landscape architects, examined the City's data and illustrated the fact that there was a park and open space deficit in the surrounding neighborhood and that the community itself was characterized by diversity and a lower average income level. These facts made it easier for the community to articulate the need for a multi-use park that could be developed by closing an adjacent street and providing more funding than was immediately available. The resulting plan included not only a preliminary layout typical of a park design project and recommendations for landscaping and storm water improvements to nearby parcels similarly formed by the skewed street network, but also an implementation strategy that included an inter- departmental property transfer, and both short and long term funding measures.

## 2.20 THE PROBLEM WITH THE RATIONAL URBAN DESIGN PROCESS

The above prototypical urban design process is a time-tested model for making rational public decisions, but there is one fundamental problem: people often do not act rationally. This results in a number of situations that hamper the pursuit of a logical course of action based on a public consensus. Hurdles to an equitable and rational decision include:

- Opposition to any change based on unsupported fear of loss.
- Poor communication – inability to understand or accept information.
- Mistrust of other groups or the agency proposing an intervention.
- Prejudicial bias against a group or type of proposed action (e.g: zoning change).
- Giving priority to one pre-selected interest over others.
- Political gain of those in power.

Many of these and other hurdles arise from the way human beings perceive and process information. As the preponderance of cognitive and social scientists have found, decisions made by human beings are influenced or determined by people's subconscious rather than the conscious, "rational" part of their mental processes. Urban designers and planners would do well to better understand how these thought process can affect individual and collective decision making. If these processes are understood, many of the hurdles described above can be reduced.

## 2.21 BRIEF OVERVIEW OF SOME OF THE MOST RELEVANT FINDINGS FROM SOCIAL SCIENCE RESEARCH

Most new cognitive science advances are based on a better understanding of the evolution, structure and functioning of the human brain. Ever since Descartes, philosophers and scientists have generally split the brain and its mental functions into two discrete components:

- The "innate", "automatic" and intuitive part responsible for regulating body functions and generating instinctual, subconscious and emotional behavior, and

- The rational, deductive part governing conscious, deliberate thought. Cognitive scientists such as Antonio Damasio have found that thetwo parts function together and that instinctive and subconscious mental activities are necessary for logical thinking and planning. Therefore, survival based, instinctual responses, deep seated values and emotions play an often dominant but underappreciated role in nearly all mental processes.

The instinctive brain's dominant role causes humans to frequently act in ways that would not be considered rational. For example, gut instincts, snap judgments and values determine behavior and opinions more than pure reasoning. And, all of us are predisposed to forming prejudices,preconceptions and group affinities.

Additionally, our automatic brain's "instinctive" reaction is to favor the familiar. As Daniel Kahnemann has demonstrated in his book, Thinking fast and Thinking Slow (2011), our decisions involving risk are typically biased toward avoiding loss rather than achieving gain. If designers and planners consider themselves change agents (and ultimately part of their mission is to help the societies they serve respond effectively to emerging challenges), then this is bad news because it means that instituting meaningful and productive changeis an up-hill effort. Most practitioners have come to realize this through experience, but surprisingly, the planning profession too often reinforce this sometimes "irrational" and counter- productive fear of loss or change. For example, just preparing an EIS that catalogues possible adverse impacts, "primes" public participants to think of a proposal in negative terms. Cognitive scientists have found that this kind of mental priming is not an insignificant effect. Planners can somewhat neutralize this tendency by focusing equally (and truthfully) on the negative effects of the "No Action" alternative or by an honest presentation of the alternatives' relative costs and benefits rather than limiting the discussion to a proposal's negative impacts. It may also be useful, when presenting a planning proposal to, for example, a city council for adoption, to tell them about the natural tendency to fear change. Note that they will likely hear a lot of fear-based criticisms of the proposal but that they will need to keep clear in their minds what the true implications of their decisions will be. This kind of warning may not carry the day, but at least it alerts the decision makers to this natural tendency.

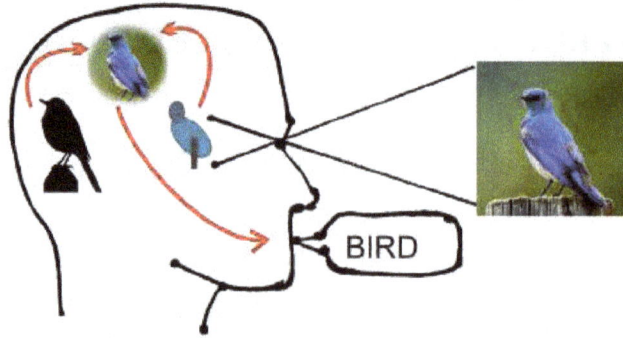

We process information and "think" by comparing "images" (either sensory or internal) to those stored in the brain. Therefore, cognitive functions are oriented toward pattern recognition, fitting new experiences into stored memories, and inductive logic.

See *The Origin of Wealth* by Eric Beinhocker and *Proust was a Neuroscientist* by Jonah Leher

**Fig. 4: The brain's process to recognize a given stimulus is to compare it to patterns imbedded in its memory. This means that the brain naturally looks for patterns and "learns" by comparing a given situation to previous experiences.**

Additionally, humans are hard wired to learn by induction; that is by drawing conclusions based on evidence, generally gathered throughexperience. This is because in order to recognizeand understand what is seen, heard, smelt orfelt, people must compare the new sensations to existing mental images lodged within their automatic brain. This basic and automaticcognitive function of comparing and resolving new and existing mental images results in humans being amazingly adept at pattern recognition (finding common elements or themes in a series of events or items) and pattern completion (filling in the gaps of information to produce a complete picture of what is being perceived). Humans "instinctively" search for causal relationships and connections between events. For the same reason, people also tend to prefer new information that fits with their existing preconceptions to the extent that they often reject outright demonstrated facts that clash with their notions based on experience. This is one reason that reiteration of scientific evidence, statistics, and extended logical argumentsare not compelling to those whose heredity, upbringing, experience, and beliefs endow them with a mental framework that is at odds with the scientific arguments. This can make it difficult to productively discuss the merits of evolutionary theory, the dangers of global warming, the economics of tax policy or the merits of more compact mixed use neighborhoods.

The brain's propensity for induction also meansthat humans find stories compelling. Thisis because the brain is constantly creating a storyline to understand the causal relationships of its surroundings. Consciousness, when you think about it, is in some sense the story that our mind creates to build a meaning out of the myriad of sensory stimuli. (Gazzaniga, 2011) For this reason people typically respond more tonarrative stories and examples than to declarative facts. (Beinhocker, 2006) Stories are a powerful communication tool, and there are many ways that both designers and planners can use themto explain ideas, propose actions and describeoutcomes. Stories told or augmented by picturesare particularly compelling.

## 2.22 LINGUISTICS

On another cognitive science front, linguists such as Noam Chomsky and Steven Pinker have searched for clues about mental processesby analyzing the structure of language. They found, for example, that the way a speaker constructs a phrase can indicate how she views her status relative her colleague or the way her mind conceives of space, time and causation. Scientists have identified the fundamental importance of conceptual metaphors as the basic building blocks of our mental processes. For example, when we conceptualize "love" we tend to think and speakin terms of a number of metaphors, including: "Love is a journey". "Love is health", "love is madness", etc. These metaphors not only help us better understand the concept of love, they "frame" the way in which we think about and act on that complex emotion. It has extended this notion of "framing" a concept through metaphorical language and shown that the use of metaphors can influence policy debate. For example, the phrase "tax relief" implies that taxes are an unfair burden and immediately "frames" the discussion in a way that favors the anti-tax advocate. While such metaphors are used in "rational" discussions, the fact that they act on our subconscious thinking mode makes them particularly powerful.

Urban designers can make use of this insight in their communication with decision makers and the public. One obvious way is to consider the metaphors they use to describe new concepts. For example, when describing the transformation of a commercial strip along a highway into a multimodal transportation spine with nodes of pedestrian oriented mixed use development, the phrase "linear community" describes a setting where people are linked to resources and attractions along a linear transit route rather than in a concentric pattern. But it also conveys the sense that the system is a real community with social connectedness and an identity.

## 2.23 SOCIAL CAPITAL

Another important field of work relating individual behavior, culture and the resulting institutional fabric is provided by Robert Putnam in his book, Bowling Alone (2000). Putnam notes that effective societies typically exhibit a high level of what he terms "social capital" – the connections among individuals, the social networks and the norms of reciprocity and trustworthiness that arise from them. Social capital is generated by formal and informal social interaction and group participation at the local level. In an earlier study documented in Making Democracy Work (1993), Putnam found that increased social capital in a society fostered effective governance. He noted that during the past few decades, there has been a dramatic decline social capital and equated that trend to concurrent erosion in Americans' trust in government, business and other individuals, conditions which threaten the nation's social and political institutions.

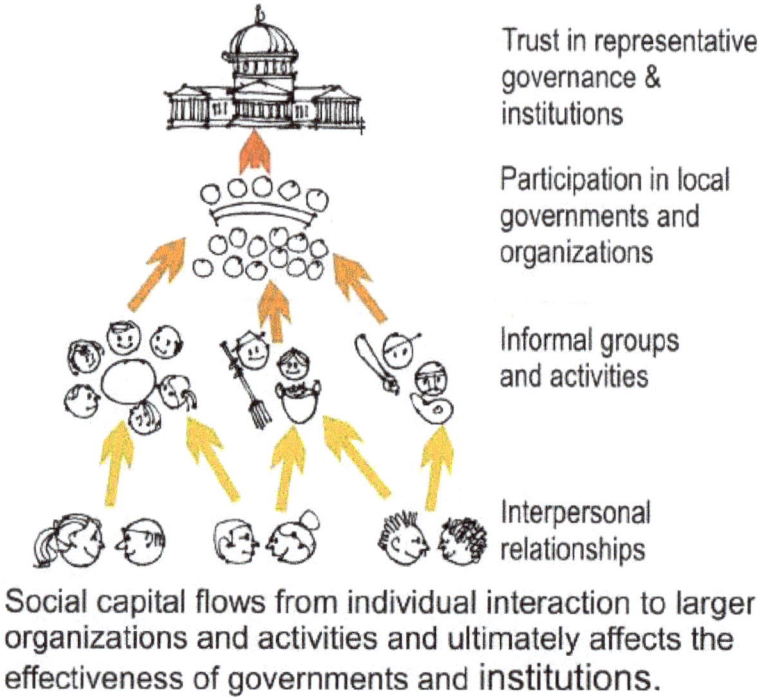

Trust in representative governance & institutions

Participation in local governments and organizations

Informal groups and activities

Interpersonal relationships

Social capital flows from individual interaction to larger organizations and activities and ultimately affects the effectiveness of governments and institutions.

See *Bowling Alone* by Robert Putnam

**Fig. 5: Social capital, that is, the connections and trust between individuals, affects people's ability to collaborate and to trust public institutions.**

If Putnam is correct, the interconnectedness of social capital is an example of individual behaviors, responding to an environmental change and creating a cultural shift that affects the nation's institutional fabric. Putnam's prescription for improving the country's social capital includes better public transportation and zoning laws, and efforts to encourage voting and political involvement. Certainly there is work for designers and planners in this agenda.

## 2.24 THE ROLE OF URBAN DESIGN

Urban designers can play an important role in the city/community building process in at least two ways:

- ◉ As a problem solving supporter of a larger comprehensive planning or infrastructure development effort, and

- ◉ As a leader or manager of a complex multi-disciplined professional team working on a complex project.

## 2.25 URBAN DESIGN IN A PROBLEM SOLVING SUPPORT ROLE

Urban designers are often called upon to assist project teams with specialized expertise in physical form giving. In these instances, urban design is often used to solve a particular problem, add a dimension to the project, or resolve a conflict. For example, Washington State's Growth Management Act (GMA) requires many of the state's cities to prepare comprehensive plansthat respond to certain objectives in the Act, including greater concentration of development (as opposed to urban expansion into rural areas), critical area protection, housing options, etc. While the GMA's focus is on comprehensive planning, the Act has resulted in quite a bit of work for urban designers who are engagedin solving the numerous planning challengespresented by the Act's implementation. Figure 22 identifies some of the GMA issues that required urban design solutions.

Urban designers are also frequently called upon to support infrastructure projects such as highway corridor improvements, bridges, port developments, and other capital improvement projects. In this case urban design can be used to enhance engineering structures with human scaled elements that enhance the project's visual appeal and/or pedestrian qualities.

### URBAN DESIGN AS A GROWTH MANAGEMENT TOOL

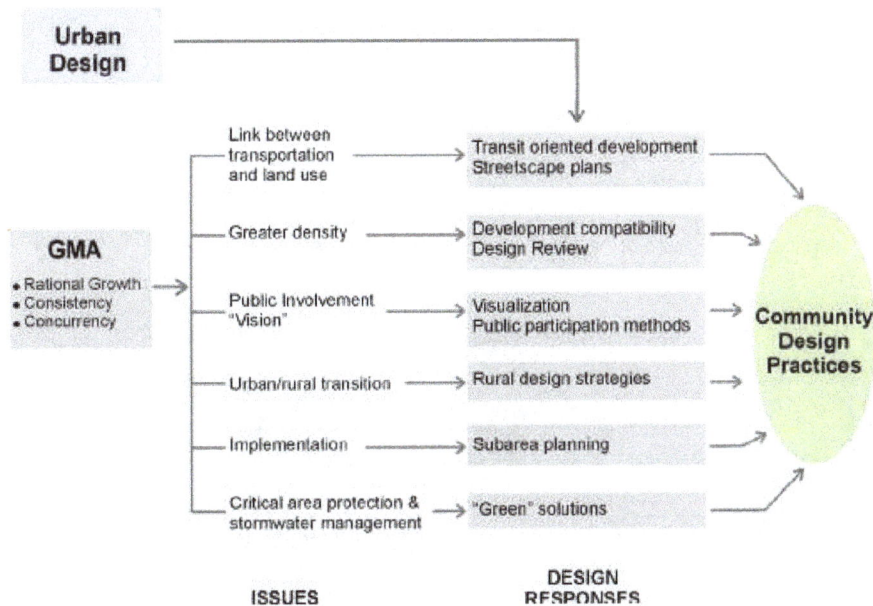

Fig. 6: Urban design solutions responding to challenges presented by growth management activities.

## 2.26 THE URBAN DESIGNER A PROJECT TEAM LEADER

Urban designers are often assigned to be leaders of multidisciplined teams engaged in complex projects because of their generalist technical backgrounds, physical design and communication tools. Because urban design touches, and hopefully integrates other disciplines including transportation, land use, environmental protection, housing, etc., it is a logical discipline around which to organize, for example, a downtown or community plan, a transit oriented development effort, or a large scale redevelopment strategy. Further, the ability to address a community's physical character through urban design exercises has proven to be needed in public engagement efforts because people tend to care strongly about the physical and visual aspects of their communities.

A leadership role requires that the urban designer assume project management duties, coordinating the various aspects of the project's process and technical elements. And, it requires some background in the other disciplines involved on the project. However, urban design, whether a person's original discipline is planning, architecture, or landscape architecture, is a good route to pursue if one is interested in project leadership.

# 3

# IMPORTANCE OF URBAN DESIGN

## 3.1 INTRODUCTION: AN OVERVIEW

- ◉ Urban design is the shaping of a community's physical form in a way that considers a multiplicity of objectives and interests through an inclusive, public decision-making process. Combining the practices of architecture, planning, and landscape architecture, urban design addresses the functional and aesthetic qualities of the physical environment at a range of scales, from the individual streetscape, park, or block to the larger community, city, or region.

- ◉ Although often thought of as being limited to intense urban settings, urban design tools and methods have been successfully used in suburban and rural communities and has even provided solutions to address environmental management challenges. Such tools have proven invaluable in implementing Washington State's Growth Management Act in many ways, such as integrating land use and transportation infrastructure, focusing an intense mix of uses in urban center and transit hubs, providing housing opportunities, achieving compatibility between existing and new development, incorporating protected natural systems (e.g., stream corridors) into the urban fabric, and enhancing the livability of a wide variety of old and new communities.

Urban design, as defined by the late University of Washington Professor Meyer Wolfe, is the "manipulation of the physical environment" in a way that:

- ◉ Addresses the way people perceive and behave in their surroundings,

- ◉ Considers the implications of form-giving actions (including the environmental and ecological consequences) at a range of scales (sometimes from the individual to the regional),

- ◉ Pursues multiple objectives for multiple clients (including affected members of the public), and

◉ Is conducted through an explicit decision-making process that offers the public the opportunity to participate in a meaningful way, identifies goals and objectives, analyzes existing conditions, explores alternate concepts and solutions, evaluates options with respect to project goals and public values, selects the preferred alternative or combines preferred elements into a synthesized concept, and includes an implementation strategy.

This is a powerful definition because:

◉ It carries a set of implicit values that all applicable urban design activities should pursue.

◉ It provides a useful checklist for designers, planners, engineers, and other practitioners to use such that they are addressing urban design's inherent values (as noted above).

◉ It describes a rational participatory process and provides a clear methodology for applying urban design concepts.

## 3.2 URBAN DESIGN PURSUES MULTIPLE OBJECTIVES FOR MULTIPLE CLIENTS

A critical, defining aspect of urban design that separates it from single-client master planning is that urban design is directed toward accomplishing a variety of objectives for all populations in a community. This sometimes involves balancing different interests, but a real benefit is that urban design can provide solutions that address more than one problem. For example, in rezoning a neighborhood to accommodate a variety of infill housing types, design guidelines can help ensure that the new structures "fit" with their neighbors.

The image above is a design for a proposed lid over the SR 520 Roanoke interchange in Seattle. This design helped satisfy multiple objectives, including addressing concerns of adjoining neighborhoods and reconnecting portions of Seattle's historic Olmsted Boulevard and its open space network. Although local transportation was the project's focus, the urban design elements were necessary to build a consensus among agencies and local residents.

## 3.3 URBAN DESIGN ADDRESSES THE SENSORY ENVIRONMENT

Urban design addresses how people perceive and use their environment. People care about the look, feel, and livability of their communities, and urban design tools are a planner's most effective tools to address this need. To accomplish this, urban designers must be well-versed in the way human perception and behavior is affected by their physical surroundings, which also involves understanding cultural behaviors and preferences, economic factors, and functional activities associated with the physical environment.

For example, the Olympia Downtown strategy, which is shown in the image below, focused on a number of actions to reinforce the design character of the downtown's six sub districts, thereby increasing its visual and functional diversity. The strategy also included a number of key elements to deal with economic development, sustainability, and social equity that were supported by the design elements.

## 3.4 URBAN DESIGN CONSIDERS THE IMPLICATIONS OF FORM-GIVING ACTIONS IN A RANGE OF SCALES

A successful urban design project typically addresses conditions within the project boundaries but also the recommendations effects on the larger surroundings. At the same time, such efforts should examine how the proposed actions relate back to the experiences of the individual. Urban design is often thought of addressing only urban design features, such as a park, street, or town centre, but urban design tools are also effective in addressing regional, landscape-scale objectives.

The Delridge Triangle Plan is a good example of this approach. As image below demonstrates, the designers consider the local socio-economic context and related opportunities (i.e., the parks walkability gap, county equity score, and locations of several unused public right-of-way) in the redevelopment of the project site. Looking at the community context helped the neighbourhood argue for additional city resources.

As Ian McHarg demonstrated decades ago, urban design methods have proven useful in addressing regional issues, as demonstrated in this downloadable graphic from *A Regional Open Space Strategy for Puget Sound*.

## 3.5 URBAN DESIGN USES AN EXPLICIT, PUBLIC DECISION-MAKING PROCESS

Broad and focused engagement techniques are critical in most public planning efforts and urban design brings with it a number of tools to help people participate meaningfully in the design process. This includes visual preference surveys in which participants evaluate different building types, park features, or environmental measures to identify which might fit best within their community. People also seem to respond well to hands-on exercises that allow them to identify the type and location of desired improvements. Children and youth also can add their thoughts through such exercises, and many urban design issues can be evaluated using web-based tools.

Urban design offers a wide variety of public engagement tools that allow for meaningful participation, and an urban designer can play an important role in the city/community building process in at least two ways:

i. As a problem-solving supporter of a larger comprehensive planning or infra-structure development effort; and

ii. As a leader or manager of a complex, multi-disciplined professional team working on a complex project.

Urban design solutions have been key in implementing growth management activities and humanizing large infrastructure projects, as the image below demonstrates.

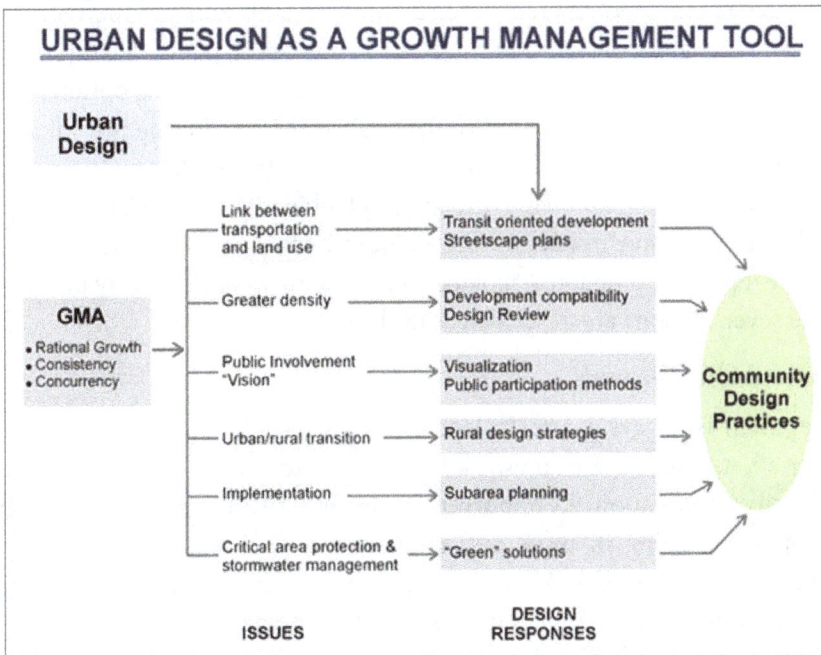

At the same time, because urban design integrates other disciplines — including transportation, land use, environmental protection, housing, etc. — it is a logical discipline around which to approach, for example, a downtown or community plan, a transit-oriented development effort, or a large-scale redevelopment strategy. Looking back at the definition of urban design has led me to the conclusion that, whether practiced by architects, landscape architects, or planners, it plays an important role in the broader spectrum of design and planning activities, and it can be a good career choice for those individuals who want to address some of our communities' most critical challenges

## 3.6 THE IMPORTANCE OF URBAN PLANNING – THE SEVEN KEY REASONS

As cities continue to grow, so do their challenges and complexities. Urban planning presents an all-encompassing solution to those challenges. Developing urban areas require a great deal of forward-thinking and thorough planning. The growth of cities and towns should happen in accordance with specific objectives that a state or territory defines, with local councils playing a significant role in it. Cities rely on urban planning to remain functional, grow in population, and attract businesses. Every crucial aspect of an urban environment is under the effect of how its layout is planned. This includes key infrastructure, transportation, and city area layout and density. Urban planning is of immense importance since over 80% of Australians are either city residents or work in urban areas, or both. The main aspect that attracts such a large number of people to Australian cities has to do with the quality of life, which is based on infrastructure and easy access to vital institutions and services, such as healthcare and education. Highly developed cities largely contribute to the economic health and productivity of society. On the other hand, poor urban planning can bring on opposite, harmful effects like constant heavy traffic, insufficient infrastructure, inadequate housing options. Such areas can also become exposed to a number of natural hazards, including fires, flooding, and climate change. Urban planning also influences property development since the city's sustainability depends on following the state-set objectives. The way our cities grow undoubtedly has a massive impact on the economy, ecology, and quality of life. And in this article, we'll provide a detailed explanation of the main reasons why urban planning matters. The seven reasons are described as under:

### 3.6.1. Planned City Growth

City development is always most efficient when it's orderly and in line with a specific vision. After all, it relies on a framework that takes into account the current and future needs of the city's population. As opportunities for work and education keep improving, cities are rapidly drawing in more and more people. For this reason, city growth is unavoidable yet predictable, making plans for future expansions a necessity. Since the core idea behind urban growth is to improve on the benefits of city living, many factors go into proper city planning. These factors include public welfare, equity, efficient emergency measures, and community participation. A planned city will take all this into consideration and transform its environment so that it can accommodate the growing number of residents. Planned city growth can create a strong relationship between local leaders, various departments, and citizens. And as a result, the entire community can pursue a common goal of productivity and better and safer living.

### 3.6.2. Improved Health and Quality of Life:

When it comes to concerns that most city dwellers share, quality of life is among the greatest. It's true that greater opportunities can attract people to urban environments. But once they settle in a city, they're faced with higher costs of living that have a significant impact on their quality of life. And as vehicles keep growing in number, city traffic also becomes a key factor for the quality of life, requiring continuous updates to traffic infrastructure and regulations. With urban planning, the concern about the quality of life will be taken into account and make the infrastructure and public spaces regulated and properly distributed. With a thoroughly planned layout, a city can provide its residents with access to all essential services, points of interest, and amenities. At the same time, the unfavorable aspects of urban life are reduced, leading to an overall healthier lifestyle and improved quality of life.

### 3.6.3. Less Environmental Impact

As the entire planet faces issues caused by global warming, cities are starting to adapt more and more to environmental considerations. In fact, the United Nations recently issued a list of development goals with the purpose of reaching sustainability, making impact assessments mandatory for development projects in cities. City planning is a key element in this regard, as it allows for proper handling of waste, a level of control over greenhouse gas emissions, and a more rational utilisation and distribution of resources. Planting trees, emphasising public transport to reduce fuel consumption, and raising public awareness about the environmental impact are all measures that could slow down and reduce the ecological harm that cities inevitably cause.

### 3.6.4. Better Economy and Resource Utilisation

Well-developed cities have sprawling markets with plenty of job opportunities. Naturally, this boosts the city economy and impacts every aspect of urban life. However, cities still need investments to keep their economies growing and they are always competing for financial input. Urban planning can ensure that more jobs are available while living costs decrease. This can happen through expanding nearby rural areas into urban environments and distributing economic activity in a coordinated fashion. Without economic considerations in city development, the living standard of city residents could be jeopardised. Today, many cities are experiencing considerable migration due to a decrease in work opportunities and rising costs of living. And the only way to prevent these negative changes and the economic decline that follows them is through proper urban planning.

### 3.6.5. National Development

Urban areas have always been the centres of development and growth, often leading the nation into progress. That situation remains the same today and the critical role cities play in national development will likely become even more pronounced.

### Why?

According to the United Nations' estimate, over 3 billion people will live in cities by the year 2050. This means that, in the near future, the impact cities will have on regional and national development will become the primary factor. As cities continue to grow, their economies, social conditions, and environmental concerns scale along. Well-planned cities will therefore become their nation's driving forces, while urban areas that lack such planning might prove highly detrimental.

### 3.6.6. Disaster Prevention and Greater Credibility

Urban planning can allow for better predictions and reactions to natural disasters. With sound strategies reinforced by the rational placement of infrastructure, cities can prevent most calamities or, at the very least, reduce their harmful effects. The better a city is planned, the more prepared it will be for any future events. The urban area will become a safer place for numerous residents, visitors, and tourists alike. As cities reach continuity regarding safety and access to crucial infrastructure, they build more credibility due to predictable, favourable conditions they nurture. And the credibility of a city reflects on its growth, economic stability, and community engagement. In light of this potential, building credibility becomes a long-term investment. Cities that reach stability in this regard find great success and disregard political changes in their plans. They remain safe and prosperous environments that attract more new residents, investments, and opportunities. Well-planned cities are also less prone to loss of property or lives and more resilient against all types of disasters.

### 3.6.7. More Efficient Problem Solving

Problem-solving doesn't necessarily mean reacting when an issue appears. Instead, a more efficient way to resolve issues is through anticipation. If a problem is anticipated, it can be secured against from the start or, in the worst case, would require less effort to solve. Sufficient urban planning allows cities to deal with potential challenges quickly and more easily through anticipation and the utilization of thought-out spatial patterns and infrastructure. This means that in a properly planned city, all areas will be made as efficient as possible and accessible for maintenance. In the event a swift reaction is needed, unhindered communication flow and readiness of city services will be precious. Through good planning, cities can decrease or remove the possibility of significant issues. And if problems arise, they can be dealt with promptly with minimal cost or damage.

## 3.7 PLANNING FOR BETTER LIFE IN URBAN AREAS

Urban planning encompasses and influences many aspects of city life. From matters of economy, through social and environmental concerns, to safety and wellbeing, a thorough, consistent plan can make life in urban areas almost idyllic. On the other hand, a lack of such plans can turn a city into an unlivable environment. In modern times, more cities are turning towards sustainable solutions that require urban planning, which affects urban development to a large extent. Archistar can be used by town planners to aid them in making optimal planning and development decisions. They can discover the highest and best use of any site, saving time whilst also bringing certainty and transparency to the development process.

# 4

# COLLABORATION BETWEEN ARCHITECTS AND PLANNERS IN AN URBAN DESIGN STUDIO

## 4.1 INTRODUCTION

Specifically, informed by the literature on the differences in design style between architects and planners and the literature on the benefits of interdisciplinary collaboration, we conceptualised the course as an experiment focused on the following research questions: Are there significant differences in design approach between the architectural and the planning students? If so, what can the two groups of students learn from each other through interdisciplinary collaboration? These questions are significant for pedagogy, since answering them enhances understanding of the specific benefits of interdisciplinary student learning. They are especially important for the pedagogy of design, since they address the crucial role of interactive learning in design studios. Furthermore, they are equally important for practice. As despite common wisdom that architecture and planning are closely related, the two fields have developed into separate cultures. By clarifying the differences between the two fields and highlighting the potential for mutual learning, research can provide a basis for mapping out more meaningful ways in which they can collaborate in practice. To address the research questions, we use four data sources: the development of the urban design projects of the student teams from the beginning to the end of the class; the students' verbal reports about their collaborative experiences, which were delivered during formal bi-weekly meetings; the results of a questionnaire distributed among all students after the course; and follow-up in-depth interviews with selected students. In short, we found that the architectural and the planning students approached the urban design problem differently. Key distinctions included different views on the importance of the relationship between individual buildings and the site, and different ways of initiating the design process - analytically or intuitively. We also found that precisely because of these differences, significant interdisciplinary learning occurred. Admittedly, the study's conclusions are exploratory, because they are based on a single case involving a small sample of participants from two neighbouring universities. Thus, the findings should serve as a basis for more systematic empirical research.

## 4.2 BENEFITS OF INTERDISCIPLINARITY

Interdisciplinarity is an approach to knowledge-generation which challenges the more common, disciplinary, approach. The disciplinary method assumes that knowledge must be acquired within the frameworks of the traditional, post-Enlightenment academic fields. It purports that meaningful, in-depth knowledge can be only generated via scientific differentiation and specialisation. Interdisciplinarity, in contrast, capitalises on connection-making between the disciplines. In this, interdisciplinarity relates to multidisciplinarity (pluradisciplinarity). However, there is a key difference between the two concepts. Multidisciplinarity typically refers to knowledge-building, which occurs when problems are addressed through the lens of several disciplines operating in parallel to each other. Results from the disciplinary examinations are then compared and contrasted. Interdisciplinarity takes a step further. It fosters learning between the disciplines and seeks their analytical and methodological integration.

"Interdisciplinarity aims at contributing to the restoration of the unity of the sciences and in the long run, of the unity of our world view." In the words "Interdisciplinary is a means of solving problems and answering questions that cannot be satisfactorily addressed using single methods or approaches."

A great inspiration for interdisciplinary collaboration, particularly in pedagogy, is provided in the works of American pragmatist John Dewey. In The Child and the Curriculum, Dewey argues that we need to "get away from the meaning of terms that is already fixed", and "see the conditions [of a dispute] from another point of view, and hence in a fresh light." This requires, he claims, "travail of thought", or to use more contemporary language, the difficult intellectual work to think self-critically, and listen to and learn from others who embrace different points of view., it further denounces the traditional separation of the disciplines as a basis for either developing theoretical knowledge or solving practical problems, and stresses the importance of an interactive, interdisciplinary curriculum in which learning occurs via conversation (with texts, peers and teachers), collaboration and constructive conflict.

Dewey's ideas echo in today's influential constructivist pedagogical approach, which also favours interdisciplinarity. This approach is grounded in the belief that knowledge is not absolute, but socially constructed, and thus cannot be passed 'down' from the expert-instructor to a passive audience of student-recipients. It espouses the so-called 'student-centred' method over more traditional and hierarchical classroom formats (e.g.

the lecture format), and advocates interdisciplinary student-to-student interaction as a means of developing independent and critical thinkers. The benefits of this approach include building mutual respect between students and between students and faculty and, in the long run, fostering greater appreciation of diverse ethical, political, gender and disciplinary views, which in turn prepares students to become more democratically minded and socially aware citizens. We focus exclusively on the process of interdisciplinary student-to-student learning, which occurred in the urban design studio.

## 4.3 DISCIPLINARY DIFFERENCES BETWEEN ARCHITECTS AND PLANNERS

- Both architects and planners are designers. Both are concerned with the arrangement, functionality and appearance of urban spaces.

- Both conduct urban design projects. In fact, the field of urban design is commonly defined as the intersection of architecture and planning.

- However, architectural and planning approaches to urban design are likely to be different because the two professions have evolved on separate trajectories through the twentieth century, at least in the USA.

- Arguably, the two have developed into separate 'subcultures'.

- The nature of these differences is a vast and complicated topic, and any brief summary will be a gross over-generalisation. Here we review only three basic, but interrelated, professional differences suggested by the literature: differences in design focus, in design decision-making, and in the value placed on the individual versus the collective contribution to design.

## 4.4 DIFFERENCES IN DESIGN FOCUS

It is hard to dispute that architect place a stronger emphasis on physical form; they prioritise the visual, the tangible, the aesthetic. Granted, there is a prominent line of architects and architectural pedagogues who have experimented with broadening this focus - from Jean-Nicolas-Louis Durand in the early 1800s to Eriel Saarinen in the 1930s and Peter Calthorpe today. Furthermore, a focus on physical design per se does not preclude a concern with broad social contexts. As the Bauhaus, and the more recent 'socially conscious' and participatory schools of architecture, have shown us, physical design can be employed for progressive social ends. Still, the traditional focus of architects, even when broader social change is at stake, has continued to be on the creation or transformation of physical form.

This is much less true for planners. Although early planning attempts to solve urban problems in the US were also centred on physical transformation, by the mid-1900s this focus had dissipated. The possibility of achieving social via physical change was severely criticised in the 1960s and a focus on physical form was viewed with disdain. After the 1960s, planning became dominated by economic and equity concerns. Its methods gravitated decisively toward those of the policy sciences (Alonso, 1986). Of course, most master plans continued to include a physical component and design courses stayed on the core curricula of the best US planning schools. Still, today's planners typically view a focus on physical forms as only one among several other foci, such as economic development or affordable housing policies.

## 4.5 DIFFERENCES IN DESIGN DECISION-MAKING

Related to the difference in focus is a basic distinction in the decision-making processes of the two professions. Wyatt (2004) puts it succinctly: faced with the same design problem, planners behave more like scientists; architects more like artists. In other words, planners use an "analytical, people-orientated, 'left-brain' approach," while architects embrace a "synoptic, theoretical, 'right-brain' stance."

If we use Schon's (1983) dichotomy of thinking styles, which differentiates between those grounded in "technical rationality" and those grounded in "intuition", or Riding and Cheema's (1991) continuum of styles for processing information and making decisions, which differentiates between the "analytists" and the "wholists", then we could conclude that planners gear to the left, while architects to the right. Typically, planners first set clearly formulated goals, then collect 'objective' data and analyse it, following an established 'scientific' model, and reach decisions only after the entire sequence of steps is complete (e.g. Levy, 2000 on comprehensive rational planning). Architects, in contrast, tend to approach a problem as an integrated whole, less empirically and sequentially, but more intuitively, introspectively and artistically.

## 4.5 DIFFERENCES IN VIEWS ON THE VALUE OF THE INDIVIDUAL VS THE COLLECTIVE CONTRIBUTION

The last principle difference we discuss here is related to the planners' and architects' views of the role of the individual vs the collaborative in design. To begin with, both professions have a somewhat troubled history of outright individualism. In planning, the early to mid-twentieth century was dominated by the grand masters, who produced visionary schemes for reform. But the failure of such expert-driven grand designs brought about humility to planning. For about 30 years, the keywords that planning students learned were not 'expert blueprints' but, rather, 'collaboration' and 'public participation'. Today, planners are seldom portrayed as solo experts, but rather as humble public servants, who inform the citizens, learn from them, and help them make their own choices. Not surprisingly, the most valued quality of US planners, as a recent study found, is communication and people-skills.

The evolution in architecture is less unidirectional. Unlike planners, architects cherish artistic creativity, a concept embedded in the broader idea of the virtue of individual freedom. Statements of legendary arrogance by Frank Lloyd Wright  Mies van der Rohe (who claimed that lay people have "no capacity to choose," Knox (1988, p.165)) or Le

Corbusier (who said that "The design of a city is too important to be left to its citizens", Scott (1998)) are examples of a long tradition which glorifies the heroic artist standing outside of society and leading the way with his/her sharpened sensitivities.8 This tradition may explain why architecture's highest honours, the Pritzker Prize, the American Institute of Architects' Gold Medals and the Rome Prize, go to an individual and not a team.

Granted, there is an important counter-tradition of collaborative and participatory architecture, which includes such important names as Ralph Erskine, Lucien Kroll and Christopher Alexander. And, many premier US architecture programmes, such as those at the University of Michigan and the Rochester Polytechnic Institute, are in the process of reshaping their studio cultures to embrace a more interdisciplinary and collaborative means of design-making. Still, evidence of a 'cult' toward the solo architect abounds. The influential Boyer and Mitgang (1996) Report, as well as the recent report of the American Institute of Architecture Students Koch et al., 2002), pointed to the rugged individualism cultivated in architectural schools as a pressing problem. This view has been echoed by many architectural educators. The Dean of the University of Minnesota's College of Design recently noted that architectural schools continue to create "star designers" proud to be "free from [the] constraints" of the surrounding social and physical context. The Dean of the University of Michigan's College of Architecture also argued that most architects continue to cherish individual authorship, and students are consistently trained as "solo artists" who use design as a "vehicle for personal exploration and expression". This approach, he noted, leads them to create signature buildings, which do not relate to the surrounding context and even negate it in order to stand out.

## 4.6 POTENTIAL FOR INTERDISCIPLINARY COLLABORATION BETWEEN ARCHITECTS AND PLANNERS

In sum, the literature suggests key disciplinary differences. In conducting the studio, we did not aim to judge which approach is 'right'. Rather, we were interested in how, if at all, these differences affect the students' design process. Would architects focus on physical form more than planners? Would the two groups use different decision-making? Are planners more open to teamwork? If so, what would each group learn from the other? To use Dewey's words, would students teach each other to see a problem "from another point of view and hence in a fresh light?"

We felt that our students would provide a good case study because they were - until the studio - immersed into curricula dominated by either profession. While in large universities like Harvard, Michigan or Pennsylvania, architecture and planning are part of the same college and interdisciplinary interaction does occur, this was not our case. The Architecture Programme at Bowling Green State University is in the College of Technology. It gives little exposure to social science courses and has no planning offerings. The Planning Programme at the University of Toledo is in the College of Liberal Arts, has a social science focus and no design studios. In the first joint class, we found that each group was unaware of basic professional concepts used by the other: e.g. the planners did not know what a figure-ground study was; the architects did not know what zoning was.

We perceived this lack of 'knowing the other' as a major learning opportunity. By facilitating the cross-mural collaboration, we aimed not only to expose students from one discipline to the logic, language and methods of the other, or merely help them acquire additional skills. Rather, we hoped to force the rethinking of deeply held assumptions of how to define problems and solve them - the type of rethinking, which Dewey identified as the major benefit of interdisciplinary learning. In the paragraphs below, we outline the specifics of our exercise, followed by our observations on the differences between the two groups of students, and our assessment of how interdisciplinary learning occurred.

## 4.7 THE STUDIO: SITE, ASSIGNMENT, COLLABORATIVE ORGANISATION AND OUTCOMES

### 4.7.1 Site

The site of the design project was a once-gracious historic plaza in the City of Toledo (Ohio), a city which is located in the immediate vicinity of both universities. In its current state, the plaza (named the Civic Center Mall) presents many problems in dire need of solutions - problems emblematic of the broader challenges facing the city and its centre.

Once a thriving industrial town, with a rich architectural heritage, Toledo has for several decades been plagued by poverty, unemployment and crime rates that exceed the national averages (poverty rates in 1999 were 18% as compared to 11% nationally and crime rates were 8060 per 100,000 as compared to 3980 nationally.

Downtown has high rates of office vacancies (19% in 2004) (CB Richard Ellis, 2004) and houses just a couple of percent of the city population, which makes it an empty shell of buildings after the close of business hours. Toledo's problems have been worsened by a notorious lack of good leadership and by a lack of cooperation, in planning and otherwise, with the surrounding wealthy suburbs. The Mall served as a microcosm of the downtown's social and physical shortcomings - from lack of planning and design coherence, to lack of meaningful land-use blend, from lack of economic activity to lack of residential diversity. It encompasses 80 years of visionary, but largely unsuccessful, planning and design efforts. The first plan, prepared in 1924, was inspired by the City Beautiful Movement.10 It proposed several buildings in the neoclassical style to frame an open Mall area, with the County Courthouse from 1897 as a terminating focal point.

However, of the planned seven, only two buildings were completed. In the mid-1940s, the local newspaper commissioned renowned architect and industrial designer Norman Bel Geddes to create a new plan that would include the Mall. Geddes' (1945) Tomorrow Plan, was based on Le Corbusier's modernist vision. However, it, too, was never realised. Subsequent proposals for the Civic Center Mall shifted the terminal focus from the Courthouse to a proposed civic auditorium. These plans were also never implemented.

Regardless of the failure of the plans, however, various new buildings were added sporadically over time. Today all of those house civic uses, most having to do with some exercise of punitive public authority (e.g. a court house, a jail and a police station). The additions occurred without much attempt to establish design coherency - something which is clearly visible in the lack of pedestrian connections between the buildings. The heritage of the City Beautiful was offset by rather plain-looking, if not dull, modernist buildings from the 1960s. The buildings do not have much aesthetic or functional relationship to each other, nor do they frame legible space. Located in a downtown with a small population, modest commerce and abundant vacant spaces, and barely connected to its surroundings, the Mall is underused most hours of the day (see Figures 1 and 2).

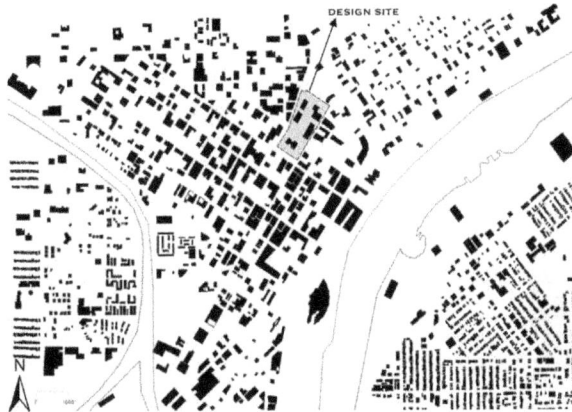

**Fig. 1: A figure-ground image of Toledo's downtown reveals its many vacant lots**

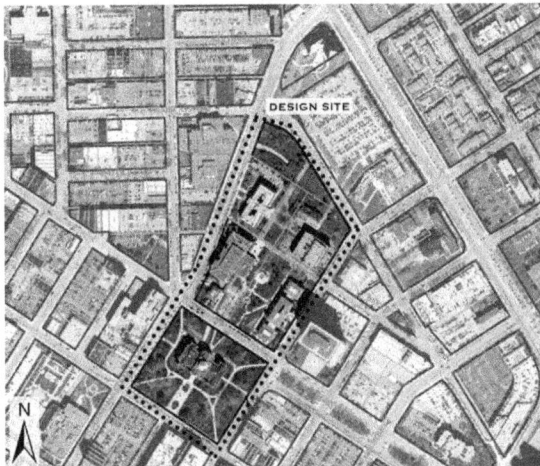

**Fig. 2:  An aerial photograph of the site**

### 4.7.2 Final Design Work

Final design work was displayed, via multi-media presentations, at the Public Library in Toledo. The forum was open to all citizens. Formal invitations were mailed to all local architects, urban planners at the City and in private practice, and housing and community development groups. At the end, it was estimated that over a hundred people attended the forum.

Five team proposals were presented: Bridging, Embracing the City, Markets, Stage and Metamorphosis. Each included a statement of goals and strategies, drawings, scale models and PowerPoint slides. Presentations lasted 20 minutes each. The teams were free to divide their time as they wished, but all teammates were required to participate in presenting their work. The presentations were followed by questions from the public and a reception. As a finale, we also produced a poster displaying all proposals (see Figure 3).

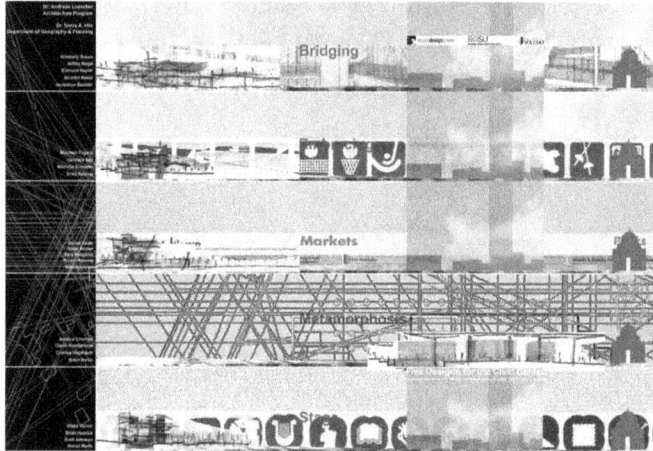

**Figure 3 A poster comprising parts of the urban design proposals of the five teams 5 Observations on the differences in urban design approach and on mutual learning**

The findings are based on our notes of how the five proposals developed (including notes taken during the students' verbal statements of how they worked and what they learned from each other); a questionnaire distributed after the class; and discussions with the members of the team, which we believed collaborated most meaningfully and produced a cohesive project. The findings are organised below in five sub-sections.

### 4.7.3 Differences in Design Focus

As we expected, the architectural students were strongly focused on physical forms. They started the project by studying the existing forms - via a photographic survey, figure-ground studies, and sketches of existing buildings - and then promptly moving to sketches of potential new structures. In all the teams, these new structures quickly became the proposals' centrepieces. This approach was questioned by some of the planners, who taught that additional analyses - of functions, users and circulation -must be performed before moving on to designing new structures, and who were not sure that new structures were even necessary. This became a source of tension in some teams. For example, one graduate student planner, who took charge of the teams' presentation of an early proposal-in-progress, spoke more about the site and neglected to articulate the details of the proposed significant new structure. This omission produced dismay among his teammates - they were concerned that if the new physical structure was not presented, the class may get the erroneous impression that they were not proposing anything at all. But from the planning point of view, "They [the architects] had their hearts set on creating new buildings from the beginning. As if without a new building, they had no project."

For the architectural students, the new structure embodied their broad vision of urban transformation - a vision which was then exported to all physical elements of the site. For example, if the theme was Embracing the City - ostensibly meaning embracing its history and diversity - the new structure was an arched glass screen that literally 'embraced' (connected) the main existing buildings. The form of the structure itself carried the central vision. It was proposed at the first in-class presentation (see Figure 4). Over time, other 'embracing' elements were integrated in the proposed amphitheatre, the existing facades and the main new site elements (e.g. benches and water features; see Figure 5).

However, this approach, which made physical form the central bearer of meaning, genuinely eluded the planners. As one observed: "Of course I am used to starting a plan with a vision. But for me vision is something practical like, say, Create Livable Downtown. It is the kind of thing that I can make into a strategy like 'build more housing', but I can't think of a way to put it into actual form. But for them form and vision are one."

### 4.7.4 Differences in Decision-Making

One common planning complaint was that the architectural students are 'too quick'. This concern reflected differences in approaching the design problem. For the architectural students, a vision for change came integrally out of the perceived problems of the site, immediately following the first couple of site visits. As one explained, "I can't remember which of us first mentioned it but I think it was right there after we walked the site. It has such potential and it is so broken down that the Embracing the City idea was kind of an obvious thing. Embracing meant bringing the place together. And once we had it, we began the design work."

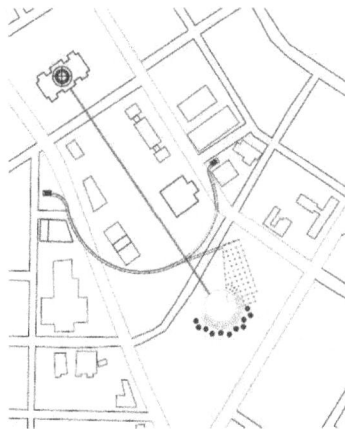

Fig. 4: Early drawing by the team working on the project Embracing the City. The 'embracing' theme was proposed by an architectural student right after the first site visit.

From there on, the 'embracing' screen carried the architectural vision of urban unity. The screen was intersected by a central pedestrian axis uniting another 'embracing' element, the amphitheatre, and the most important historic building on the site, the Courthouse.

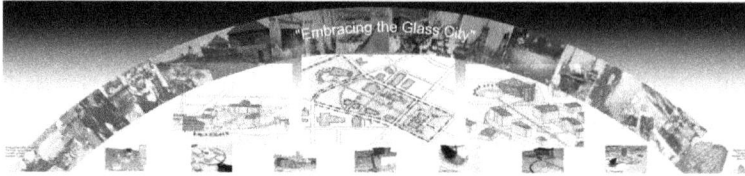

Fig. 5: 'Embracing' elements were carried on in all other main physical elements, and even in the design of the presentation poster. Notably, the poster showed details of the site design in seven boxes, each representing one of the criteria for good urban design according to William Whyte's The Social Life of Small Urban Spaces. Whyte's theory was introduced to the architects by the team's planner. It proved instrumental in that it provided the team with a logical framework, and helped it articulate goals and strategies.

But for planners, such a quick movement from problem to solution was foreign: "They [my teammates] start with sketching and playing with the site. I don't think they first think about it as I am used to - what's the history, what are the functions, who lives and works nearby and who visits. I think I try to follow logical steps from beginning to end. I guess this is engrained with me.

They work by immediately modelling the physical solution. I can appreciate their boldness. I apparently don't have it! For me, we hadn't yet figured out what the problem is and they already had the solution!"

### 4.7.5 Differences on Collaboration

While the literature suggests that architects are less open to collaborative work than planners, we did not observe signs of such a difference and did not receive complaints that any architect was ignoring his or her teammates. It may be that such a difference does exist. Our study, however, was not well designed to capture it, since by emphasising collaboration to begin with, we likely suppressed any student's impulse to display individualistic behaviour. This is a limitation which we address further in the conclusion.

Rather, we observed that the different views on the balance of the individual versus the collaborative role were reflected in the design of the proposed structures - the architects preferred that the new structures stand out as individual signature pieces; the planners wished to make these structures conform to their surroundings.

To begin with, in all five teams, the initial buildings proposed by architects stood out by their size - all were larger than the Mall's crown jewel, the Courthouse. In one case, the new building was larger than all existing ones combined. This produced dismay in both instructors and in the planning students. Eventually, all new structures were substantially scaled down.

Disputes also emerged regarding style. Architects were interested in innovation and radical visual contrast between the proposed and the existing; planners in emulation and stylistic cohesiveness. One planner proposed as a project motto a quote from Daniel Burnham's Group Plan for Cleveland, in which Burnham eloquently praised unity of style over individuality. The idea was quickly shot down by the teams' architects as too restrictive. The planner also suggested design guidelines, which would ensure that the new buildings echo key stylistic elements of the historic buildings on site. These were eventually accepted in a watered-down version by the team. The planner explained: "I thought the new building should compliment what is already on the site, maybe not literally but in principle, in some subtle or modified form. And then I suggested using Burnham's quote as a motto. But this idea did not have much appeal for them. They wanted a very large, contemporary building. They thought contrast would work better, strengthen the site more than would consistency. They kept pointing that harmony does not mean same style or similarity, that 'complementing' does not mean 'emulating'. And they used Gehry's Art Museum's addition as an example and said, 'Wouldn't it have been terrible if he just replicated the old museum? No, luckily he complemented it by bringing a new idea. So maybe they are right. Or maybe the truth is in the middle - I guess that's where we ended up!"

## 4.8 EMBRACING THE CITY: AN EXAMPLE OF INTERDISCIPLINARY LEARNING

Below we present the progress of the Embracing the City proposal and some aspects of interdisciplinary learning, which occurred in the process. The team comprised four students: one planner from the University of Toledo, who was a practicing professional, and three senior undergraduate architects from Bowling Green State University.

The architecture students began the project by drawing sketches and creating models and photomontages. As previously noted, their idea of embracing part of the Mall with a glass screen was born early on (see Figure 4). They felt the screen would connect the disjointed fragments of the site and carry a strong symbolic message. The planning student forced rethinking of the project by introducing theory on what constitutes good urban spaces, which she thought should be discussed, prior to proposing any design solutions. Specifically, she presented her teammates with a summary of recommendations for successful place-making based on William Whyte's book *The Social Life of Small Urban Spaces* (1980). She critiqued the size of the screen and the fact that while it connected some of the buildings, it passed right in front of other buildings, thus dividing rather than embracing the space. She questioned the time spent designing the screen while neglecting simple site problems, such as the lack of seating and a coherent pedestrian trail system. And she demanded that a cohesive written statement of the site's problems be produced, prior to proceeding with design. Eventually, the screen became only one of the project components. Other elements included brief guidelines for future development, a proposal for re-landscaping and re-paving the site, and renderings of how some of the existing facades should be redone. The screen's footprint was redefined so that it no longer divided the space. The screen structure became lighter and was lifted above ground in several places to allow free pedestrian flow. It was also adorned with photos from Toledo's history. The project proposed a new amphitheatre and a mixed-use building opposite the Courthouse, as well as the conversion of some existing buildings to commercial and residential use, in order to create multi-use space. In this team, as in the others, the planner steered the team's attention from the new building to the site. But her key contribution was to aid the architects in developing methods of formal reasoning. Prior to her intervention, the architects were inclined to work from instinct and focus on form-building. By introducing Whyte's criteria for good place-making, the planner brought logical substance to decision-making. She helped the team clarify its goals and develop a framework for evaluating which design ideas may and which may not work. This framework ultimately formed the skeleton of the team's final work and presentation (see Figures 5 and 6). As two of the architects put it:

"I think discussing the book [by William Whyte] and generally talking to her [the planner], was the most helpful thing because it made it possible for us to talk about the things which we wanted to accomplish. We, of course, knew most of these [Whyte's] principles before we started - they are kind of common sense points that should be part of any good urban design – but seeing them on paper brought it all together. It helped the project move along because now we kind of knew more clearly what we are aiming at, what makes sense and what doesn't, and could explain it to others."

"I think before the studio, I was used to being given an assignment; say, 'Design a gymnasium'... But I never came up with the actual assignment. In other words, somebody had already decided that a gymnasium was necessary - so somebody had already given me a solution to a problem (e.g. the problem of not having recreational opportunities) and I had to only refine it. But in the studio, my teammates and I had to actually go through a process of deciding what is needed for the site, so we had to ask ourselves questions and come up for ourselves with what the problems and the solutions are before anything else. And I think she [the planner] had more experience in this."

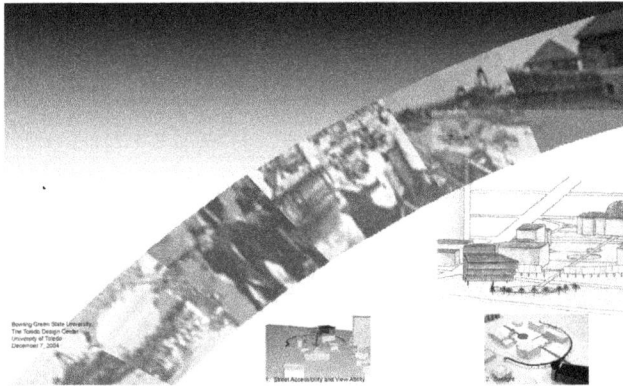

**Fig. 6:  A detail of the same poster showing site elements, which ostensibly applied two of Whyte's urban design principles: street accessibility and viewability, and sunlight**

## 4.9 STUDENT COMMENTS ON INTERDISCIPLINARY COLLABORATION

A questionnaire distributed after class rendered the following results: 16% of the respondents thought the studio was very helpful, and 60% thought it was helpful in developing their interaction skills; 8% rated the interdisciplinary collaboration as excellent, and 68% rated it as very good (see Table 1).

Free written comments also showed that most students appreciated the potential of interdisciplinary collaboration to enhance learning. Representative remarks include: "The collaboration was a useful forum for exchanging ideas, learning from one another, and helping to prevent a tunnel vision on the part of the architecture and [on the part of] the urban planning students. Much can be learned by cross-training students in this way." "It's always beneficial for students to collaborate with others, especially outside of their departments and disciplines. I believe this combined experience was worthwhile and with a little tweaking could become a real asset to Toledo, the two universities, and an annual event at the Design Center."

**Table 1. Students' evaluation of the learning outcomes from the interdisciplinary studio (based on their responses to select questions from a questionnaire filled in after the class)**

| Was the experience helpful to you in: | Very Helpful | Helpful | Not so helpful | Of no significant value |
|---|---|---|---|---|
| Developing collaboration/ interaction skills | 16% | 60% | 16% | 8% |
| Developing a critical perspective | | 60% | 32% | 8% |
| Developing public speaking/ presentation skills | 16% | 52% | 8% | 24% |
| How would you rate the experience in terms of: | Excellent | Very good | Good | Inadequate |

| The quantity of skills gained from interaction | 8% | 60% | 32% | |
|---|---|---|---|---|
| The quality of skills developed from interaction | | 68% | 32% | |
| Overall evaluation of interdisciplinary interaction | 8% | 68% | 24% | |

## 4.10 FUTURE RESEARCH IN INTERDISCIPLINARY URBAN DESIGN

The studio pursued two main goals: pedagogical (to enhance students' learning by exposing them to interdisciplinary teamwork) and research (to conduct an experiment on the differences between architects and planners along three key axes: design focus, design decision-making, and views of the individual and the collaborative role in design). As noted above, we encountered a logical difficulty regarding the third axis. We could not effectively judge whether the architects were less inclined to work collaboratively than the planners, since we made collaboration an explicit requirement to begin with. While this is a limitation of the study, we felt that had we treated our students purely as research subjects (had we not required intensive teamwork), we would have failed our pedagogical responsibility. To correct for this deficiency, we suggest that future efforts to measure differences in interdisciplinary studios include surveys and interviews, not only at the end but also at the start of class (i.e. before proceeding with interdisciplinary teamwork).

Putting this limitation aside, the experiment showed that disciplinary differences do exist. Indeed, the architects did place greater emphasis on physical form and approached the problem more intuitively than the planners. They also cherished design pieces which would stand out from the rest of the site - a finding which adds fuel to Kelbaugh's (2004) and Fisher's (2000) views of the high value assigned by architects to 'signature' pieces. Furthermore, the study illustrated that while differences exist, substantial mutual learning may occur via serious interactive 18 S. Hirt and A. Luescher work. To use Dewey's words, we observed that students underwent a process of "questioning of entrenched beliefs and positions," which allowed them to "get away from the meaning of terms that [are] already fixed". The process by which the architects in the Embracing the City team moved, with the aid of the planner, from an intuitive grasp of the situation to a logical framework is a good example.

Given the need for interdisciplinary teamwork in solving complex urban problems (Sebastian, 2003), future research must elaborate on the differences in values, logic and methods between architects and planners, since misunderstanding these differences hampers real-life collaboration (Wyatt, 2004). In order to better document the differences, a future three-semester-long study, for example, could first pose an urban design problem to architecture students, then to planning students, and lastly to interdisciplinary teams.

In doing so, such a study will highlight the disciplinary differences in their 'pure form', and will also show how values, logic and methods evolve via interdisciplinary interaction.

Finally, we recommend that US programmes look to enrich their design curricula. Many European schools (e.g. the Eindhoven Institute of Technology) include courses in interdisciplinary and collaborative design. Such courses must enter US schools as well, if the 'great divide' between architects and planners is to be ever bridged.

# 5

# URBAN DESIGN, SOCIALIZATION AND QUALITY OF LIFE

## 5.1 INTRODUCTION

Architecture as a discipline is essentially integrative: connecting past, present and future, drawing on art, science and the social sciences, balancing qualitative with quantitative factors. Good architecture and urban **Trouble Shooting** design contribute to making cities both functional and attractive to residents and visitors. While architecture is about the design of buildings, urban design is about the relationships between the buildings, the roads and spaces that they front, and the people who make use of them. The outstanding building projects are those that are not only visually stimulating, but are also sensitive and respectful of their surrounding developments and environment. A well-designed city takes into consideration this important relationship between buildings and the beauty of the city as a whole.

*The process of socializing or sociability in a city means acquiring the model of style life of that city. Administrators of the municipal cultural and social realm can reinforce the suitable models of citizens' social behavior and learning by cultural and social planning to improve the culture of urbanization. This process should be done considering the basic needs of the zones and neighborhoods. From the other viewpoint, socializing is a state that all of the society members have to learn the urban life style and get ready to be known as formal citizens of the society. This style is habits, costumes and the way of living in the city that the administrators of the city have to teach it to the citizens in the form of acquisitive patterns of behavior. The municipal cultural and social realm has to develop the way of acquiring the necessary life skills gradually among the citizens*

Socialization is the process that prepares humans to function in social life. It should be re-iterated here that socialization is culturally relative - people in different cultures are socialized differently. The process of socializing or sociability in a city means acquiring the model of style life of that city. The process of socializing includes every daily activities of citizens' life. People have mutual relationship in this place and actualize it through presenting the municipal cultural and social activities and resolving the needs of each other. The municipal cultural and social realm has to develop the way of acquiring the necessary life skills gradually among the citizens. These patterns can be taught through an effective and mutual relationship, this education should be presented to acquire the necessary abilities considering the cultural values and norms of a society continually and constantly.

The nature and conceptualizations of public space and public life have been always associated with collective participation and socialization, in other words, with the capacity to live together among strangers. Today these associations seem to have become challenged and problematic, and often end in questioning whether public space still matters for our public life? This uncertainty has become somehow evident in the rising scholarly interest in the last two decades debates on the future of our cities' public life and public spaces. However, most of the research produced has been extremely limited and narrow in its scope and often taken disparate positions especially between academics and practitioners, above all it has been dominated by narratives of profound loss and lament. Thus, it has failed to provide an understanding of the new context of social change we are in at the moment. In this paper we define urban design, socialization of citizen and relation between them.

## 5.2 URBAN DESIGN

### 5.2.1 The art of Creating and Shaping Cities and Towns

Urban design describes the physical features that define the character or image of a street, neighbourhood, community, or the city as a whole. Urban design is the visual and sensory relationship between people and the built environment. The built environment includes not only buildings and streets, but also the natural environment such as shorelines, canyons, mesas, and parks as they shape and are incorporated into the urban framework. Urban Design is a process to foster quality in the built and natural environment.

Urban design involves the arrangement and design of buildings, public spaces, transport systems, services, and amenities. Urban design is the process of giving form, shape, and character to groups of buildings, to whole neighbourhoods, and the city. It is a framework that orders the elements into a network of streets, squares, and blocks. Urban design blends architecture, landscape architecture, and city planning together to make urban areas functional and attractive.

Urban design is about making connections between people and places, movement and urban form, nature and the built fabric. Urban design draws together the many strands of place-making, environmental stewardship, social equity and economic viability into the creation of places with distinct beauty and identity. Urban design is derived from but transcends planning and transportation policy, architectural design, development economics, engineering and landscape. It draws these and other strands together creating a vision for an area and then deploying the resources and skills needed to bring the vision to life.

Urban design and city building are surely among the most auspicious endeavours of this or any age, giving rise to a vision of life, art, artifact and culture that outlives its authors. It is the gift of its designers and makers to the future. Urban design is essentially an ethical endeavor, inspired by the vision of public art and architecture and reified by the science of construction." -Donald Watson. Urban design involves place-making - the creation of a setting that imparts a sense of place to an area. This process is achieved by establishing identifiable neighborhoods, unique architecture, aesthetically pleasing public places and vistas, identifiable landmarks and focal points, and a human element established by compatible scales of development and ongoing public stewardship. Other key elements of place making include: lively commercial centers, mixed-use development with ground-floor retail uses, human-scale and context-sensitive design; safe and attractive public areas; image-making; and decorative elements in the public realm.

Urban design practice areas range in scale from small public spaces or streets to neighborhoods, citywide systems, or whole regions. Good architecture and urban design contribute to making cities both functional and attractive to residents and visitors. While architecture is about the design of buildings, urban design is about the relationships between the buildings, the roads and spaces that they front, and the people who make use of them.

The outstanding building projects are those that are not only visually stimulating, but are also sensitive and respectful of their surrounding developments and environment. A well-designed city takes into consideration this important relationship between buildings and the beauty of the city as a whole.

Context of the surroundings, the socio-cultural and physical climate, which determines the building form, scale and timing of developments Connectivity, which should be comfortable and convenient for both people and vehicles: and Contribution to the streetscape, greenery, the public spaces and the community.

### 5.2.2 General Urban Design Goals

- Provide a built environment that respects natural environment and climate.
- Improve the quality of life through safe and secure neighbourhoods and public places.
- Use pattern and scale of development to provide visual diversity, choice of lifestyle and social interaction.
- Foster the continuation of districts, communities, and neighborhoods as

distinguishable Subareas within the city.

⦿ Create activity centers as places where people gather and interact.

⦿ Maintain historic resources as important landmarks that maintain the city's historic identity.

⦿ Utilize landscape as an important aesthetic and unifying element.

# 5.3 SOCIALIZATION

## 5.3.1 Elements of Socialization

Socialization is a fundamental sociological concept, comprising a number of elements. While not every sociologist will agree which elements are the most important, or even how to define some of the elements of socialization, the elements outlined below should help clarify what is meant by socialization.

## 5.3.2 Socialization: Becoming Who We Are:

The structural-functionalist perspective points out that the fundamental task of any society is to reproduce itself—to create members whose behaviors, desires, and goals correspond to those that the particular society deems appropriate and desirable. Through the powerful and ubiquitous process of socialization, the needs of society become the needs of the individual. Socialization is a process of learning. To socialize someone is to train that person to behave appropriately. It is the means by which people acquire a vast array of social skills, such as driving a car, converting fractions into decimals, speaking the language correctly, or using a fork instead of a knife to eat peas. But socialization is also the way we learn how to perceive our world; how to interact with others; what it means to be male or female; how, when, why, and with whom to be sexual; what we should and shouldn't do to and for others under certain circumstances; what our society defines as moral and immoral; and so on. Although socialization occurs throughout our lives, the basic, formative instruction of life occurs early on. Young children must be taught the fundamental values, knowledge, and beliefs of their culture. Some of the socialization that occurs during childhood—often called anticipatory socialization—is the primary means by which young individuals acquire the values and orientations found in the statuses they will likely enter in the future (Merton, 1957). Household chores, a childhood job, sports, dance lessons, dating, and many other types of experiences give youngsters an opportunity to rehearse for the kinds of roles that await them in adulthood.

## 5.3.3 Goals of Socialization:

Arnett, in presenting a new theoretical understanding of socialization (see below), outlined what he believes to be the three goals of socialization:

    i. Impulse control and the development of a conscience

    ii. Role preparation and performance, including occupational roles, gender roles, and roles in institutions such as marriage and parenthood

iii. The cultivation of sources of meaning, or what is important, valued, and to be lived for In short, socialization is the process that prepares humans to function in social life. It should be re-iterated here that socialization is culturally relative - people in different cultures are socialized differently. This distinction does not and should not inherently force an evaluative judgement. Socialization, because it is the adoption of culture, is going to be different in every culture. Socialization, as both process or an outcome, is not better or worse in any particular culture.

It should also be noted that, while socialization is a key sociological process in the development of individuals who can function in human society, not every aspect of human behavior is learned. For instance, there is evidence that most children have innate empathy for individuals who are willfully injured and consider it wrong. Thus, some aspects of human behavior that one might believe are learned, like empathy and morals, may, in fact, be biologically determined. To what extent human behavior is biologically determined vs. learned is still an open question in the study of human behavior.

### 5.3.4 Importance of The Process of Socialization in Our Life:

The importance of socialization in our life can hardly be exaggerated. The following description makes it very clear.

- ◉ **Socialization converts man, the biological being into man, the social being.** Man is not born social; He becomes social by virtue of the process of socialization. Various instances likethat of Kaspar Hauser, Anna, the wolf children of India and others have made it very clear that only through constant training the newborn child becomes social in nature.

- ◉ **Socialization contributes to the development of personality.** Personality is a product of society. In the absence of groups or society, no man can develop a personality of his own. But socialization is a process through which the personality of the new born child is shaped and molded. Through the process, the child learns an approved way of social life. At the same time, it also provides enough scope for the individual to develop his individuality.

- ◉ **Helps to became disciplined.** Socialization is social learning. Social learning is essentially the learning of rules of social behavior. It is the values, ideals, aims and objectives of life and the means of attaining them. Socialization disciplines an individual and helps him to live according to the social expectations.

- ◉ **Helps to enact different roles.** Every individual has to enact different roles in his life. Every role is woven around norms and is associated with different attitudes. The process of socialization assists an individual not only to learn the norms associated with roles but also to develop appropriate attitudes to enact those roles.

- **Provides the knowledge of skills.** Socialization is a way of training the newborn individual in certain skills, which are required to lead a normal social life. These skills help the individual to play economic, professional, educational, religious and political roles in his latter life. In primitive societies for, example, imparting skills to the younger generation in specific occupations was an important aspect of socialization.

- **Helps to develop right aspiration in life.** Every individual may have his own aspirations; ambitions and desires in life. All these aspirations may not always be in consonance with the social interests. Some of them even be opposed to the communal interests. But through the process of socialization an individual learns to develop those aspirations. Which are complementary to the interests of society. Socialization helps him to direct or channelize his whole energy for the realization of those aspirations.

- **Contributes to the stability of the social order.** It is through the process of socialization that every new generation is trained acceding to the Cultural goals, ideals, and expectations of a society. It assures the cultural continuity of the society. At the same time, it provides enough scope for variety and new achievements. Every new generation need not start its social life a fresh. It can conveniently rely on the earlier generation and follow in cultural traditions. In this regard, socialization contributes to the stability of the social order.

- **Helps to reduce social distance.** Socialization reduces social distance arid brings people together if proper attention is given to it. By giving proper training and guidance to the children during their early years, it is possible to reduce the social distance between people of different castes, races, regions, religions and professions.

- **Provides scope for building the bright future.** Socialization is one of the powerful instruments of changing the destiny of mankind. It is through the process of socialization that a society can produce a generation of its expectations can be altered significantly. the improvement of socialization offers one of the greatest possibilities for the future alteration of human nature and human society.

### 5.3.5 Relation Between Urban Design and Socialization:

- The process of socializing or sociability in a city means acquiring the model of style life of that city. Administrators of the municipal cultural and social realm can reinforce the suitable models of citizens' social behavior and learning by cultural and social planning to improve the culture of urbanization. This process should be done considering the basic needs of the zones and neighborhoods.

- The process of socializing includes every daily activity of citizens' life. People have mutual relationship in this place and actualize it through presenting the municipal cultural and social activities and resolving the needs of each other. So, the municipal cultural and social realm have to present the special cultural identity of that city directly or indirectly to develop cultural evolution. From the other viewpoint, socializing is a state that all of the society members have to learn the urban life style and get ready to be known as formal citizens of the society. This style is habits, costumes and the way of living in the city that the administrators of the city have to teach it to the citizens in the form of acquisitive patterns of behaviour.

- The municipal cultural and social realm has to develop the way of acquiring the necessary life skills gradually among the citizens. These patterns can be taught through an effective and mutual relationship, this education should be presented to acquire the necessary abilities considering the cultural values and norms of a society continually and constantly.

- The family as the central core of urban society that has an important role in educating the individuals for adjusting to current urban life style can have an effective reflection in promotion of citizenship cultural improvement in establishing such processes. Administrators of cultural and social realm of the city can move toward the development of socializing by presenting cultural and social programs for families especially for women, children, teenagers, youths and the aged.

- The school as the second social institution in promotion of citizenship improvement culture is a very effective element. The municipal office cooperating with education office can play a very effective role in development of citizenship education and improvement of the amount of citizenship culture. From the view point of sports, the development of urban health, promotion of religious culture development and so on, is one of the aspects of cultural and social activities that the municipal and education offices can do cooperating with each other.

- Mass media can play an important role in socializing the citizens by devoting the educational cultural and social programs, producing series and fiction movies, presenting cultural and social messages and also attractive and interesting contents about solving the urban problems.

- Mosques are one of the essential institutions that can effect on acculturation and socializing of citizens. Mosques as social and cultural institutions can have an effective function in solving the problems and citizenship educations in all quarters. The citizens in all quarters can be encouraged in the process of social learning by giving the necessary education.

- The cultural institutions in the city are one of the essential institutions in citizens' socializing, however they have different functions and they can present an important role in development of urban sociability cooperating with the municipal cultural and social realm.

◉ The municipal executive administrations in cultural and social realm with considering the aspect of socializing should pay attention to behavioral system and the patterns of mutual relationship among citizens in the form of teaching of social norms .These patterns can be presented in a descriptive and directly/indirectly way and can help to solve and improve the urban culture. For example presenting the patterns of (traffic dos and don'ts and the right culture of driving) can help in learning the urban discipline very much by institutionalizing the patterns of urban behavior in citizens' minds.

◉ These behavioral patterns should present in the form of habits and the traditional and Islamic ways that are accepted by the society along with reinforcement of social costumes that observing them is necessary in the city. Urban laws that are the other aspects of these behavioral patterns can be effective in institutionalizing and teaching the citizens.

◉ Social values in the city as the element of social system of the city have the effective function in a desirable way in supervising the citizens and directing them toward a transcendental life. These elements should be promoted through assessment, rated and classified in the positive and correct form like (teaching books and reading them, respecting to observing the citizenship rights, observing urban laws, the form of contact and interaction with each other, respecting to the citizens, paying attention to beautifying of urban furniture, using the land space in the best way and so on) are parts of the value system of the city. Administrators in the cultural and social realm of the municipal management should promote the norm system of common values of citizens considering the sources and value criterions of the society like scarcity and basic needs and develop the social consensus and preserving and respecting to them in the way of conceptualization and in the form of identified along with public feelings and judgment.

# 6

# THE PUBLIC REALM AS
# THE ARMATURE OF CIVIC LIFE

## 6.1 INTRODUCTION

Urban design practice at this level is focused on solutions to problems of urban sustainability and community resilience in the face of future change. For architects who are familiar with urban design practice, the information presented here will likely be well understood, but its didactic presentation here it may still offer some additional insights. For those in our profession who have chosen different types of practice, this step-by-step guide is intended to provide insight into the particular techniques and priorities that inform urban design as its own, unique discipline.

**Simplified structure of urban design practice that covers a full range of scales:** master planning, block and street design, and infill urban projects. It is easiest to think about this practice as being constructed around a set of project objectives and a series of design strategies.

**Objectives:** Most urban design practice is constructed around four main objectives:

   i. Walkability – To promote public health and respond to consumer preference in terms of livability preferences
   ii. Multi-modal Mobility Options – To increase personal choices and decrease carbon footprint
   iii. Mixed-use or Multi-use Development – To provide market flexibility and to support choices of sustainable urban/suburban lifestyles.
   iv. Ecological Awareness – To understand and enhance the role of nature in an urban environment Strategies:

Overall objectives generate a series of major urban design strategies to create an infrastructure of public spaces that is functional, safe, aesthetically pleasing, commercially successful, well-connected and accessible to diverse populations. This primary strategy relies heavily on the following three sub-strategies for successful resolution:

- ⊙ To design buildings as the walls to urban rooms
- ⊙ To create a particular sense of "Place" from the generic medium of urban space
- ⊙ To Create a high-quality pedestrian environment

## 6.2 THE RELEVANCE OF URBAN DESIGN IN A VIRTUAL WORLD

Today, many cities are envisioning a future that is ever more connected and "smart." In this digital context, the word "smart" is defined in many different ways, but perhaps the simplest way of thinking about a "smart city" is one that has digital technology embedded across all city functions. A smart city uses its integrated communications technologies (ICTs) to fuel sustainable economic and physical development by managing three critical areas:

### 1. Traditional Physical Infrastructure

- ⊙ Transport
- ⊙ Energy / utilities
- ⊙ Public safety
- ⊙ Environmental protection and enhancement

### 2. Civic Governance

- ⊙ Administrative services for citizens
- ⊙ Cultivating civic engagement through participatory and direct democracy

### 3. Economic Development

- ⊙ Promoting innovation in industries, clusters and districts of the city
- ⊙ Supporting "knowledge-intensive" companies and investments
- ⊙ Supporting a workforce geared to the "knowledge economy" through good quality education.

In such highly connected digital environments, many transactions take place in the virtual world rather than in physical space, and this has led some urban theorists to postulate that physical places are now less important than they were in traditional cities. An extension of this argument would suggest that urban design is a matter of choice, not necessity. However, the rebuttal to this "place-less" argument posits that physical places, and all the meanings embedded in them, are more important now than they ever have been. In this context, urban design is a vital endeavour.

While digital technology keeps pushing us apart, using media to bridge physical distance, we as a culture continue to gather in specific locations that are meaningful to us. The smartest places, therefore, are those that combine the best of both the physical and virtual worlds, where presence and "tele-presence" are fused together at a location (Mitchell, 1999. 143). Here, attractive and sustainable physical locations are penetrated by information and communication technologies to provide a collaborative meshing of physical and virtual environments, with both local and global dimensions. In this way the centrifugal forces of technology are balanced by centripetal ones of human interaction in physical space. And the design of this real, physical space is crucial (Walters, 2011).

To this end, it's possible to list some of the features of smart community planning and design ranging from municipal policies to planning strategies to detailed urban design concepts. These concepts can be briefly summarized as:

◉ Promoting diverse, compact and mixed-use neighbourhoods that are walkable and transit supportive

◉ Defined and accessible public spaces, both urban and natural

◉ Energy efficient buildings that follow the premise of "long life, loose fit," making themselves adaptable to changing patterns of use without major disruption to their external form or their urban surroundings.

These principles of public space design provide a measure of stability and consistency that are the partners in the physical world to the changes and displacements of the virtual world.

In a society that enables us to live and work anywhere we like, the places we choose to inhabit become all the more precious and important. The hyper-connected global economy, far from being placeless, needs very specific "territorial insertions". "Territorial insertions" is academic jargon for well-designed urban places, and it's apparent that as "traditional locational imperatives weaken, we will gravitate to settings that offer particular cultural, scenic, and climatic attractions -- those unique qualities that cannot be pumped through a wire -- together with those face-to-face interactions we care most about.

Put another way, when we can live and work anywhere we choose, we select places that please, support and nurture us on several levels. Virtual media are vital elements for places to be successful but they themselves are no respecters of location.

Left to their own devices, one tele-serviced spot is as good as another, with convenience perhaps the only moderator. For locations to become places in the more meaningful sense, to hold some special status in our cultural hierarchy, they have to combine the convenience of global linkage in the virtual realm with characterful physical presence, and that comes chiefly through the quality of urban design (see Fig. 1).

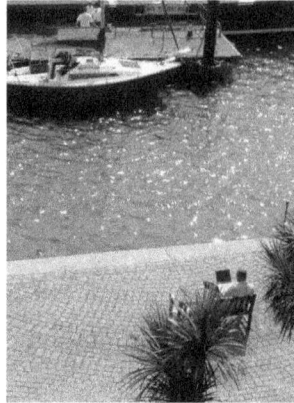

**Fig. 1: Working from laptop in a beautiful place.: Global reach, local beauty. Dartmouth, Devon, UK. Photo. David Walters**

## 6.3 OBJECTIVES OF GOOD URBAN DESIGN

Urban design is an expansive subject area, but at its core reside four key objectives that focus on urban resilience in the face of potentially serious future changes. Within the remit of physical design, the capacity of a community to respond effectively to these changes – in climate and in socio-economic circumstances – rests largely on four main considerations. These are all interlinked, but may be summarized as:

- ⦿ Walkability
- ⦿ Access to choices for personal mobility
- ⦿ Mixed-use or multi-use development
- ⦿ Ecological enhancement.

Walkability is linked today with clear consumer preferences in the property market, but at a deeper level it is linked with vital public health issues, where physical activity and a healthy lifestyle can become part of everyday life. This means having the ability to walk or bike safely and conveniently to stores, restaurants, places of worship, schools, parks, and to other transportation options—buses, trams, light rail, and commuter rail in addition to automobiles.

Walkability and cycling are also connected with mixed or multi-use development. These development patterns provide useful and attractive destinations for walks or bike rides, and real estate development markets show distinct consumer preferences for living and working in urban (and suburban) environments that are walkable. Two findings in particular from the 2013 National Association of Realtors® Community Preference Survey illustrate the importance of a mixture of uses, walkability, and connected urban space.

When asked to choose between a neighborhood that "has a mix of houses and stores and other businesses that are easy to walk to" versus a neighborhood that "has houses only and you have to drive to stores and other businesses," the walkable neighborhood was preferred 60 percent to 35 percent. Most striking in the survey is that the preference for a mixed-use walkable neighborhood is strongest for those who are in the real estate market now.

And the higher preference among those under age 40 for walkable communities, revitalizing cities, and alternative transportation tells us the high importance these community traits will have with the consumers of tomorrow. These market preferences signal a major change from decades of suburban single- use environments. This integration of uses within a community satisfies current and projected market conditions and provides the potential for health benefits; but even more importantly the more trips that can be made conveniently without car travel, the lower will be a community's carbon footprint.

The natural environment and ecology of a community is protected and enhanced by the inclusion of parks, greenways, playgrounds and ball fields; these green spaces should be integrated into the fabric of all communities at the master plan level. Beyond the benefits of walkability, studies have shown that living close to parks raises property values, and that being in parks can improve mental health, simply through contact with nature. While large athletic complexes are likely to be a drive away, neighborhood parks, ball fields and greenways can be readily incorporated into master plans. Ideally, no one living in a community should be more than a 5-10 minute walk away from some type of natural space (see Fig. 2). To quote from just one of many the academic and medical studies on this topic:

This research shows that the percentage of green space in people's living environment has a positive association with the perceived general health of residents. Green space seems to be more than just a luxury and consequently the development of green space should be allocated a more central position in spatial planning policy.

**Fig. 2: Latta Park, Dilworth, Charlotte, NC. This park, laid out by the Olmsted Brothers in 1911-12, sits within a mixed-use neighborhood, adjacent to an elementary school, and a few minutes' walk from a light rail station. The park is lined along its perimeter by residential streets, and homes facing onto the park.**

In addition to human needs, master plans also need to consider the movement of wildlife, and seek to connect as many elements of the green infrastructure as possible so as to provide safe and nurturing corridors for birds and animals.

In successful urban design projects all these more generally applicable factors are customized to suit site and community-specific objectives developed through locally based analyses. These local data sets comprise the analyses of site and social contexts, local and regional cultures, environmental imperatives, market trends, development programs and priorities, stakeholder personalities, public sentiments, and (inevitably) local politics. Through the analysis process, this mix of factors gives rise to particular local objectives, and the design team can fine-tune their approach to embrace both general and specific objectives. It is this mixture that makes each urban design project unique and challenging.

## 6.4 STRATEGIES AND TACTICS OF GOOD URBAN DESIGN

Each project develops its own unique calibration of relevant objectives, and this mix of factors can be best addressed through a combination of four main strategies and their related tactics:

- ◉ Creating an infrastructure of public spaces that is functional, aesthetically pleasing, commercially successful, well-connected and accessible to diverse populations. One useful way of developing this strategy is through the design concepts of connectivity, choice, and identity.

- ◉ Designing buildings to be the walls to urban rooms the relevant design elements here are frontages, façades, fronts, backs, and thresholds.

- ◉ Creating a sense of "place" from urban space there is a specific design vocabulary that assists this process: enclosure, extension, continuity and contrast.

- ◉ Creating a high-quality pedestrian environment. For this task, there is one key rule to follow the "Golden Triangle" of Good Urbanism.

## 6.5 INFRASTRUCTURE

Connectivity, Choice, and Identity Nothing is more important in urban design than the creation of an infrastructure of connected public spaces that are attractive, efficient, and safe. By this we mean the interconnected system of boulevards, avenues, streets, alleys, urban squares, plazas, parks, pocket parks, playgrounds, greenways and other types of public space that may be appropriate to a particular location or culture. At the scale of community master planning, this spatial infrastructure is of primary importance, almost the first thing the urban designer creates or enhances. At the more local scale of the block, urban square, or street, the design task is to reinforce or improve this spatial infrastructure; This is best achieved by the appropriate placement and design of the buildings that enclose or define the spaces "urban rooms" in the community. At the scale of infill urban architecture, the designer's task becomes one of elaboration, enhancing the sense of place by the scale, materiality, and pedestrian-scaled detail of the façades that form these urban rooms.

The key words in all of this are "connectivity" and "enclosure." Connectivity can be best defined as the density of connections in a public space system, whether it comprises any combination of busy boulevards and avenues, quiet neighbourhood streets and alleys, or landscaped greenways and trails. Connections are important because as connectivity increases, travel distances decrease, and route options and personal choices multiply. This combination of functionality, economy, and choice relates directly to two fundamental principles of public space design that focus on streets in urban areas:

- ◉ All urban streets should be multi-functional, i.e., safe and attractive for pedestrians and cyclists as well as for car, and transit where appropriate.

- ◉ Streets should connect to form a network with multiple choices of route. This connectivity spreads out traffic more evenly and reduces congestion. Both these design attributes correlate with the "Complete Streets" policies enacted by many American municipalities to ensure safe travel for people of all ages and abilities.

In this context, it's hard to overstate the importance of connecting streets into a network in any project. Connected networks increase mobility for vehicles, bicyclists, and pedestrians, and decrease costs of civic services by having more choices of routes around any neighbourhood or district. This same flexibility increases the efficiency of a wide range of city services – from public transit and school buses, to emergency police, fire and ambulance services, and garbage collection.

Street connectivity can even lead to improved water pressure and easier maintenance of the underground pipes because of the ability to loop lines through a development rather than creating dead-ends in cul-desacs. Street systems either maximize connectivity or frustrate it. North American neighborhoods built prior to 1950 were rich in connectivity, but with the advent of ubiquitous automobile ownership in subsequent decades, traffic engineers discarded this "old" idea that worked for all modes of travel, and replaced it with "dendritic" systems that focused only on the movement of vehicles. As the Canadian urbanist Patrick Condon has explained (Condon, 2010), streets in a dendritic system all branch out from the main "trunk," which in U.S. cities is usually a freeway or major state highway. Attached to this main trunk are the major "branches," which are the suburban multi-lane arterial streets or thoroughfares. These large branches then give access to the next category of the "tree," the "minor branches," which are the collector streets. Collector streets then connect to the "twigs and branch tips" of the system, the residential streets and dead-end cul-de-sacs. In a typical residential subdivision, the dendritic system requires fewer linear feet of road length per standard-sized house lot, and as such it has become universally popular with developers, eager to reduce costs. However, it has major disadvantages: almost all trips are made longer than they would be if the system were interconnected and it is prone to extensive congestion problems since the system provides no alternative routes. All traffic has to funnel into a few main intersections, which, due to very high traffic volumes are often slow to clear and always dangerous to pedestrians and cyclists.

Most importantly in today's environment where pollution and climate change are major issues, studies show that dendritic configurations force residents to drive over 40% more than residents in older, traditionally connected neighbourhoods. This results in a 40% increase in greenhouse gas emissions per car, and because these street systems are hostile to cyclists and pedestrians, households are likely to own two or more cars. Therefore the greenhouse gas emissions per household in a dendritic subdivision are easily double that of residents of older, connected districts. This is a real problem, but it's one that can be minimized by designing the alternative: a well-connected public street infrastructure. At the outset of this discussion, the reader would have noticed that we used specific names for specific types of public space, and there's a good reason for that.

"Open space" is very hard to design. What are the criteria? How big should it be? The term "green space" is similarly vague and imprecise. Urban designers try hard (not always successfully!) to avoid using these terms. Naming a space as a particular type immediately puts it in context, moves it from the general to the specific, assists in the creation of identity, and begins to suggest programmatic requirements and sizes. This of course, is exactly the same process as designing a building; the building program names and specifies various types of space - labs, conference rooms, classrooms etc. - and from these spatial types' architects can construe more detailed information regarding content, relationships, hierarchies and adjacencies. This specificity of naming spaces leads us to our second controlling design concept: spatial Enclosure. Whether streets, squares, plazas, or parks, urban designers should consider all exterior public spaces as a series of "urban rooms." These exterior rooms help create identity – being in a place -- and can be designed in the same way as the interior spaces of buildings. Understanding this analogy leads us to our second of urban design strategies: organizing buildings to function as the walls to urban rooms.

## 6.7 URBAN ROOMS: FRONTAGES, FACADES, FRONTS, BACKS, AND THRESHOLDS

The analogy between exterior spaces in the city and interior spaces in a house was first promulgated early in the Renaissance by the great architectural theorist Leon Battista Alberti, who wrote: "a city is like a great house; a house is like a small city; cannot the varied parts of the house... be considered miniature buildings?".

In this way of thinking a city's streets are analogous to a house's corridors and hallways - some grand and formal, others smaller and serving minor functions. A city's plazas are the main public rooms in the house. City parks are analogous to front and back gardens – each with different characters. Understanding this analogy allows the urban designer to position buildings to create a hierarchy of public spaces and the connections between them, where each space caters to its functionality, but is also part of a larger system. The ability to use buildings successfully as the main compositional devices in making urban rooms depends on understanding five related concepts: Frontages, Façades, Fronts, Backs, and Thresholds. Frontages have two components:

Private frontages are the areas between building façades and the rights-of-way lines that divide private from public property (on a typical street, this might be the back edge of the sidewalk).

Public frontages are those parts of public space and streetscape in front of buildings that are dedicated primarily to pedestrian use. The combination of the private frontage and the public streetscape defines the character of the public realm and constitutes the transitional layer between fully public space and the private realm of buildings.

This zone can range in character from urban to rural and frontages can vary from the formal – arcades and colonnades in high urban areas, to more relaxed front yards and porches on residential streets. Façades are defined as the external walls of buildings that front public space. These building walls define the edges of public space (they are the walls to the urban rooms) and have a special responsibility in the urban design process. As Robert Venturi so clearly stated (see section 2.4 above) façades have to respond both to the internal pressures of the program and to the external forces and responsibilities of their context. In particular, buildings have a very special responsibility at the lower stories of their façades where they interact directly with pedestrians in public space.

Fronts and Backs are self-explanatory – the fronts and rears of buildings – but this simplicity begets a universal urban design rule that should rarely, if ever be broken: Fronts face Fronts and Backs face Backs. This is a cardinal rule for successful public space and conforms to the simplest hierarchy of urban space design. Fronts relate to more formal public space and, usually, higher intensity uses, and the architecture should reflect that importance. Backs frame more private spaces. Uses are more relaxed and informal, and the architecture can follow suit.

This illustrates the principle of context-responsive architectural design. The nature of public space varies from front to back, and the architecture should reflect, embrace and support this difference. The ability for architecture to be radically different from front to back is well illustrated by one of the greatest urban compositions: the Royal Crescent at Bath, England. This speculative housing development, in reality no more than a collection of 30 town homes, was designed by John Wood the Younger and built between 1767 and 1774. The magnificent setting of the curved sweep of town homes – facing a preserved landscaped lawn and overlooking the medieval town of Bath – provided Wood with the opportunity to "pull out all the stops" in his façade design. His great curved façade features a uniform row of 114 attached Ionic columns, each 30 inches in diameter and 47 feet tall, topped by a uniform entablature, 5 feet deep (see Fig.3).

This design raises the scale of the row of town homes to one of a major urban "palace," a brilliant real estate development tactic and one that responded very well to the expansive scale of the landscaped setting.

**Fig. 3: The Royal Crescent, Bath, UK. 1767-74. John Wood the Younger, architect. The front façades of thirty town homes are reconceptualised as a unified composition at the scale of a royal palace – a brilliant real estate development strategy married with excellent design.**

By contrast, the backs are a higgledy-piggledy collection of ad-hoc bays, projections, and window styles (see Fig.4). The informal backs have evolved over time with additions and adaptive reuse projects, but the magnificent front has remained unaltered. In this way, the building responds to its two different contexts in ways that accentuate these differences and thus enhances the clarity of the urban spaces it helps define.

**Fig. 4: The backs of the Royal Crescent, Bath. Compare with Figure 3. The rear façades show how many decades of additions and conversions have left their mark of building form, volume and detail, with little regard for architectural unity.**

**Thresholds comprise two things at once:** They are links and separators. In urban contexts, a threshold is a zone of passage or pause between two spaces of different characters. Most often, this marks the transition between public and private, clearly manifest in American domestic architecture by the front porch. Traditional front porches are miniature rooms in their own right, privately occupied but fully open to public view. They are partially separated from the public realm by being raised a couple feet above the front yard and defined by an encompassing roof, columns and open railings. As a transition between public urban space and private domestic space, they are perfect examples of a threshold that links and separates at the same time.

This same concept can be scaled up to work in a fully urban context – such as Newbury Street in Boston, MA. Built (like Bath) as developer-driven town homes, this part of Back Bay has evolved into one of America's great shopping streets. In particular, the zone of transition between the public street and the buildings is a crucial element in the making of a memorable place. The change of levels – down to a semi-public courtyard and up a half flight of steps to the building entrance – creates a three-dimensional threshold that's full of visual interest, activity, and visual clues that explain elements of the urban setting (see Fig. 5).

**Fig. 5: Newbury Street, Boston, MA. Photo. David Walters**

## 6.8 SENSE OF PLACE: ENCLOSURE, EXTENSION, CONTINUITY AND CONTRAST

We have described the differences between "space," as a generic medium, and "place" as its special and more emotive variant. Places are effectively containers of memory and meaning; indeed, the clearest definition of "place" is" space enriched by the assignment of meaning" (Pocock and Hudson, 1978). In urban contexts, places in this special design sense are a function of enclosure – creating a "here" as opposed to other "theres" beyond. An easy way to think about this is to remind ourselves that "place" is experienced from within. We say we are "in" a place.

There is a fancy word to describe our attachment to places: topophilia. We are all topophiliacs. We all have a predisposition to invest locations with attachments, and good urban design can facilitate and encourage this process.

"Placemaking" connects directly with the concept of "urban rooms," and making memorable places is one of the primary purposes of urban design from the human perspective. The urban design process deals primarily with the physical design of space; it is the urban designer's task to create attractive "containers" that support and encourage a range of human activities. These intersecting patterns and rhythms of activity, and the multiple interactions they can generate between people and space, become the medium through which meanings may be generated and "places" created.

In the same way that being in a room in a building is enhanced by being able to see out, beyond the confines of that particular space, having views to other spaces, other urban rooms is a valuable element of placemaking. Seeing beyond one's present location in an urban space, being able to look from "here" into another location, a "there," provides a context for the experience of any particular urban place. This urban design concept is called "serial vision," and it was developed by the great British urban designer Gordon Cullen in his book "Townscape," published in 1961.

This premise is very simple, but of great value to the urban designer. Cullen defines these simple concepts of "here," where one occupies urban space -- on a street, in a square, at a sidewalk café, for instance -- and "there," a glimpsed urban vignette that offers other possibilities of human activities. This allows the urban designer to set up a framework of urban experience through spatial sequences of stillness, movement and progression (see Fig.6).

**Fig. 6: Via Santa Chiara, Assisi, Italy. From the shaded enclosure of the street – "here" – a view is framed through the arch to "there," the next space in the sequence at the start of the wider Corso Guiseppe Mazzini. The buildings along the Corso are angled to close the view, but in a way that creates a "deflected vista," suggesting yet further spaces to be discovered. These principles of spatial design transcend beautiful Italian hilltowns and are applicable in most urban settings. Photo. David Walters**

This process of understanding an urban area through the orchestrated experience of moving through   series of urban rooms is an excellent way to make places that are memorable and which stick in one's memory. Whatever the size or scale of an urban space, there are three basic requirements to help spaces transform in peoples' minds to memorable "places."

### 6.8.1 Placemaking Requires Enclosure

This influences the siting and orientation of buildings and the relationship between the heights of buildings to the width of the spaces between them. To generate a feeling of reasonable enclosure, ratios between 1:2 and 1:4 (1 unit of building height to 2-4 units of spatial width) provide good rules of thumb. At ratio above 1:6, all feeling of enclosure dissipates, and the space generally fails to register as a place where we would want to visit or linger. Tighter enclosures, say 1:1 for intimate streets, or 2:1 for urban lanes or alleys are also part of the urban designer's spatial palette. The ways in which streets and pathways enter and exit the space is crucial. As noted above these are the "doorways" between urban rooms. The enclosing walls can be organized to facilitate long views for clarity and a desire to impress a sense of grandeur to the location, or they can be used to obscure views in ways that promote discovery as spaces are revealed through movement.

### 6.8.2 Placemaking Benefits from a Balance Between

Continuity and Contrast. In practice, this balance can be hard to achieve. Many architects, this author included, were trained to be the "formgivers" for society, that is, the creative artists/professionals who could create new and original forms for buildings. This approach meant that modesty and respect for context were rarely given high priority in teaching or in practice. We see this trend alive and well today in the global practice of shape-bending "starchitects." Pick up any architecture magazine to see the latest example. Some of these unique buildings are indeed beautiful objects, and some do gain in visual import by being seen in contrast to their context. Frank Gehry's original Guggenheim Museum in Bilbao, Spain is a case in point. Here the shimmering metallic shards, bulbs, and curves of the museum resonate against the hard, repetitive and rectangular masonry block structure of its urban setting. This contrast works because it is singular. Being the only freeform object in a restrained rectangular context provides exceptional power. However, this power is weakened every time another zig-zag or blobby building is introduced into the mix. The more freeform objects inhabit the city, the quicker the city spatial structure dissolves. And as the city structure dissolves, so too does any sense of "place. "The moral here is simple: only a small percentage of buildings deserve the special privilege of contrasting with their contexts.

### 6.8.3 Placemaking Requires Active Edges to Space

This is a design principle that is often not fully appreciated. While the physical proportions of the urban space are important, so too is the treatment of the edges of the space, where the buildings frame the pedestrian experience. The most important thing to do here is to activate the edges, and this is best achieved by providing spaces that people can occupy, for rest, leisure, business and other unscripted (peaceful) activities.One useful concept for the design of building façades in this location is the notion of "thick edges," such as the medieval shopping colonnades along the High Street, in Totnes, England, this author's home town (see Fig. 7). or other more modest examples such as projecting stoops, awnings or balconies, or even simply recessed doorways that create shelter and shadow (see Fig.8). Properly designed tree canopy from street trees can also help make the edges of urban space into places where people want to hang out.

**Fig. 7: The "Butterwalk," medieval shopping colonnade along the High Street in Totnes, U.K. Originally designed to cover market stalls, this use continues today for retail and restaurant activity, providing very rich and active edges to the public space.**

Guidance for making places for human activity and creating memorable urban vignettes can be best summarized by the last section of this unit. Taking our cue from the precept established by the great Danish urbanist Jan Gehl, we can build on his potent statement that "Life takes place on foot" 2011. By this Gehl means that we construct the most complete understanding of urban locations through interacting with them as pedestrians. In this way we have the time and opportunity to discover, recognize, remember, and enjoy aspects of our environment that are otherwise inaccessible to us as drivers passing by at 30-50 miles an hour. This process of identification with our urban surroundings, the creation of memories linked to places, can be facilitated by a combination of architectural and urban design working in harmony. These two disciplines come together in the zone at the base of buildings where the vertical external wall meets the horizontal plane of the pedestrian realm. This sets the scene for one of the most important lessons that this unit can teach: How to create the "Golden Triangle" of Good Urbanism.

## 6.9 PEDESTRIAN ENVIRONMENT

The Golden Triangle of Good Urbanism The illustration that accompanies the last section of this course unit (see Fig. 3.8) illustrates a street in Ann Arbor, MI, and it contains most of the ingredients required for active public space. These can be summarized under two headings, the vertical plane and the horizontal plane, and gives rise to the phrase "Golden Triangle." Imagine a line vertically up the building façade for about 20-25 feet, a horizontal line extending out 20-30 feet (from the façade line to the outer edge of the layer of on-street parking) and then join these two extremities by the hypotenuse. Within that triangle, you have defined the most important zone of space for placemaking.

**Fig. 8: Street in Ann Arbor, Michigan. A good illustration of the "Golden Triangle." Photo. Craig Lewis. Used with permission.**

This triangle can then be analyzed relative to the character of its vertical and horizontal planes:

## 6.9.1. Vertical Plane of the Façade

- ◉ Lots of windows and doors at sidewalk level. This allows for visual and functional connections between inside and out and activates the sidewalk. The façade at this level is largely transparent. People look into shop windows; shopkeepers look out at the street and provide informal supervision of public space through "eyes on the street." People enter and exit doorways and mingle with other pedestrians in what Jane Jacobs referred to as the "sidewalk ballet" (Jacobs, 110).

- ◉ Inset doorways, canopies and awnings. Apart from providing shade and shadow, these indentions and projections create a "thick edge" to public space and act as useful threshold spaces.

- ◉ Vertical proportions of the buildings that line the street. Horizontal lines encourage the eye to skim quickly along the surfaces of the buildings. Vertical features and proportions establish a clear rhythm in perspective that holds the eye. This slows down the act of viewing and allows our minds to linger, process, and retain information. This aids memorability and thus the creation of memorable places.

- ◉ Lots of colored signage. This provides information and adds visual interest. The variety of signs is important; too much architectural regimentation into dimensionally reductive strips is a poor substitute for this lively mélange.

- ⊙ Interesting architectural detail on the second andsometimes the third floor of the building façades. Cornices, moldings, projections and patterning all capture and hold the eye, enriching the visual memory of the place. Pedestrians tend to absorb architectural detail up to the first couple stories only, and then again at roof or cornice level.

- ⊙ Façade levels in between this "base" and this "top" condition are less important in creating a sense of a memorable place. It is thus worthwhile to concentrate resources in these two locations, with priority always given to the pedestrian level.

- ⊙ Proportions of the ground floor façade are important. The ground floor story height should be higher than those above to create a "base" for visual stability in the overall façade. Commercial spaces in mixed-use buildings, of course, generally provide taller ground floor spaces, and this works well relative to the lower floor-to-floor heights of apartments above.

In apartment buildings this is rarely the case, so it is important to raise ground floors above the sidewalk level to create this extra height on the bottom façade story. This also provides better visual privacy inside the rooms, and also a sense of security engendered by this raised level.

The common developer ploy of slab-on-grade construction should be resisted with all possible vigor and determination. Not only does this make ground floor apartments very vulnerable to breaches of privacy and security, but it also means that the lowest floor-to-floor dimension on the facade is the same as all the ones above. This results in poor proportions, as there is no sense of the building having a visual "base." As an even more cautionary tale, this sidewalk edge should never be formed by blank walls and grills to podium parking levels. This is a sure-fire recipe for killing street life and creating the worst kind of memories in the minds of passers-by (see Fig. 3.9 for a terrible example).

**Fig. 9: Public space killed by architecture. Blank walls, "cars behind bars," and utility boxes all conspire to create a dead zone directly opposite a light rail station where pedestrian activity should be actively encouraged. It's a cautionary tale that this miserable design was authored by a fully licensed architect. Photo. David Walters**

## 6.9.2. Horizontal Plane of the Sidewalk / Street

   a.  *Dimensions:* This is the most critical factor in making an active and attractive street scene and a lively edge to the public space of the long, skinny urban room we call a street. For an active and functional pedestrian sidewalk on a commercial or mixed-use street, 12 feet is the minimum useful dimension; 14 or 16 feet are better. These larger dimensions allow several functions to take place simultaneously: walking by, stopping to chat with friends and neighbours, window-shopping, sidewalk dining, retail displays and sidewalk sales, and people entering and leaving buildings through multiple doors. As noted earlier, this all adds up to an everyday example of Jane Jacobs' "sidewalk ballet."

   b.  *On-Street Parking:* This is an essential element for active pedestrian street life. While drivers hope to snag a convenient spot close to their destination (and this visible parking enhances the viability of street level retai) the main purpose of the row of parking is to protect pedestrians on the sidewalk from moving vehicles directly next to them. The 7 or 8 foot zone for parking (plus bike lane if applicable) makes sidewalk dining and other retail /leisure activities along the edge of the street safe and attractive.

   c.  *Street Trees:* These are vital elements of any streetscape, both visually and functionally. The species and placement of the trees, along with the specification of their tree wells or grates (or landscaped swales/planting strips on residential streets) is critical. Unfortunately, there is a lot of misinformation provided by forestry specialists that leads designers astray concerning urban trees. Some urban forester advocates will insist that trees must be planted in unobstructed landscape beds large enough to encompass their drip lines, but everyday observation of some of America's best urban streets shows hundreds of examples of handsome and grand street trees, flourishing in a largely paved urban environment. Most urban trees do fine even though they may not reach their full potential growth compared to forest or field locations. In practice, commercial streets may have tree wells 5 feet square or smaller, while residential streets may have swales up to 6-8 feet wide and of indefinite length. Trees in most locations should be planted approximately 30 feet on center, and along a main commercial street, tree species should be taller and more vertically proportioned, so that when they're mature, their lowest branches are 12-16 feet above the sidewalk and don't block business signs. On a primarily residential street, the trees can be lower and spread more broadly because the buildings (mostly houses) are set further back from the street. The lowest branches only need to be a bit above the head height of a tall person.

   d.  *Lighting:* Some of the best sidewalk lighting for safety and aesthetic enjoyment comes from shopfronts themselves. This should be supplemented by regularly spaced, pedestrian scaled light fittings.

   e.  *Surfaces and Street Furniture:* These are items that get a lot of attention, as well-intentioned efforts to improve the visual quality of the public space. However, when a space is actively used and enjoyed, the human activity provides much more visual stimulation and enjoyment than brick paving patterns or other expensive fixed items. It is much better to spend that money on wider sidewalks, moveable seating, or more street trees. This unit has provided a great deal of detailed, practice-based information about urban design objectives, strategies and tactics. We hope that even those architects who have experience in this field have found it useful and that architects from other branches of our multi-faceted profession have found in instructional. The last unit in this course builds on this practice foundation to show where urban designers can have the most impact in shaping America's urban environment: the process of master planning communities. This practice has three basic elements:

- Public design charrettes to involve the community in shaping their future.

- Master plans that develop detailed alternative growth scenarios before settling on the most appropriate.

- Form-based Codes -- the zoning documents that are directly tied to the content of the master plan, and which guide implementation of the plan over many years.

# 7

# THE URBAN PLANNING CONCEPT BASED ON SMART CITY APPROACH

## 7.1 INTRODUCTION

The rapid development of human population poses various challenges in various aspects of life. Department of Population of the Division of Social and Economic Affairs of the United Nations 2017, reported that the number of current population is nearly 7.6 billion and it will increase to be 8.6 billion by 2030, 9.8 billion in 2050 and 11.2 billion in the year of 2100. The current urbanization rate is 32 percent; and it is predicted that 30.53 million population of the city is emerge annually. The increasing rates number of urban population compared to the rural one is an indicator of dynamic distribution of population around the world. Today, the occupancy of urban areas is estimated 3 percent of the planet's surface. It consumes 75 percent of global primary energy and impacted on the increase of 50-60 percent of the world's total greenhouse gases.

*Smart City is considered as a strategy to reduce the problem due to the rapid urban growth and urbanization. From the review of previous reported studies, it is resolved that smart city has provided various different perspectives. In order to understand smart city from the context of urban planning, it is essential to recognize the considerable aspects which is important from both sides.*

Urbanization phenomenon has encouraged the emergence of needs for a better quality of life. People move from one country to another looking for better living opportunities which impacted on generating the problems of high pollution, traffic congestion, waste and social problems. Other problems are the scarcity of energy resources that related to the need of healthy living, daily activities, social facilities and urban infrastructures. Almost all the problem resulted on the change in economic situation which entails to the fluctuated cost price. In addition, the increase of carbon emission has contributed the negative impact on the intensive climate change.

One of the strategic solutions which is currently being discussed to solve the above problems is the application of smart city concept into urban planning. The recent advances of Information and Communication Technology (ICT) are considered align with the need of reducing technology costs by applying low-cost mobility, free social media, cloud computing, and effective cost in handling big-data management. A model of diagrammatical ideas is proposed summarizing the understanding of smart city concept in the context of urban planning.

## 7.2 DEFINING 'SMART' CITIES

Broadband network developments (DSL, cable, satellite and wireless communication) are greatly affecting the interaction potential of various actors (e.g. individuals, small businesses, institutions and local governments,) by providing access to both worldwide knowledge and information (re)sources as well as a broad range of tools to connect both locally and globally. Based on the challenging new network opportunities, and on steering competitiveness gains and community development efforts, the concept of *'smart' communities/cities* has appeared. Searching the literature available, however, a clear-cut definition of 'smart' communities/cities does not exist. Furthermore, a number of terms similar to 'smart' communities have appeared: 'wired' communities, 'broadband' communities, 'digital' communities, 'networked' communities, 'smart community network' and 'community informatics', 'intelligent' communities; these seem to be used interchangeably by the various researchers, but all imply communities that are making *'a conscious effort to understand and engage in a world that is increasingly connected'* (Albert et al. 2009:8). Although there are certain differences in the way the above terms are used by the various researchers, all definitions have three key aspects in common, namely: the *communication mean* (network infrastructure – technology – ICTs); the *process* (networking of various actors); and the *goal* pursued (public involvement or other).

## 7.3 COMMON CHARACTERISTICS OF A SMART CITY

The most common characteristics of a smart city are:

- ⊙ The infrastructure of urban networks facilitates political efficiency, social, and cultural development.

- ⊙ Stressing on business-led urban development and creative activities.

- ⊙ Social inclusion of various urban populations and social capital integrated in urban development.

- ⊙ The sustainability of natural environment as a strategic component for the future.

The features of smart cities are formulated in various perspectives which are determined by the context. Based on the above review, the re-conceptualization of smart city in the context of urban planning covers the multi-tier architecture of a digital city planning. The city planning includes: E-Government services, E-democracy services, E-Business services, E-health and tele-care services, E-learning services, E-Security services, Environmental services, Intelligent Transportation, Communication services. The city planning is further based on sustainable environment vision to solve the local development problems on the dimensions of people, economy, governance, mobility, environment and quality of life.

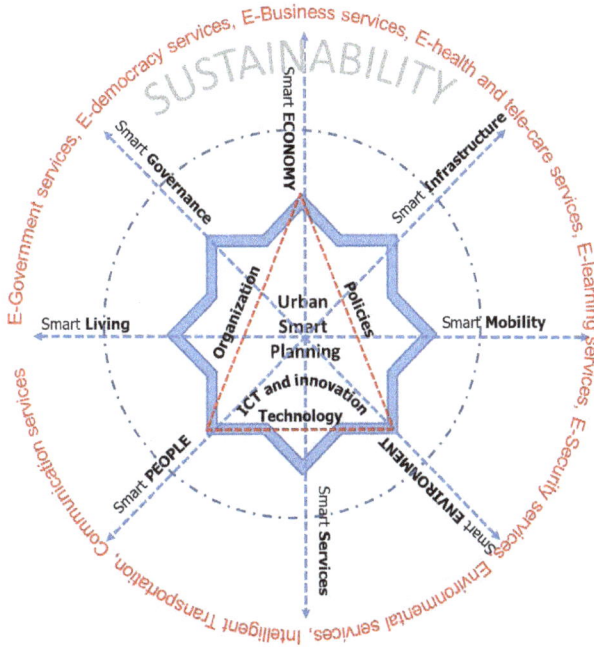

**Fig. 1: The concept of smart city in the context of urban planning**

## 7.4 SMART' COMMUNITIES

'Smart' communities are defined by the Canadian Federal Government (CFG) (2002) as those communities in which local leaders and stakeholders, by use of electronic networks and the Internet, are forming alliances and partnerships in order to innovate and extract new economic and social value. In this definition, emphasis is placed on the network deployment (transport and ICTs), but also on investments in human and social capital in support of sustainable community objectives and quality of life, by means of engaging social participation as well as user-specific technologies and community-building applications. The idea of a 'wired' city as the main development model and of 'connectivity' as a source of growth is brought to the forefront for increasing local prosperity and competitiveness. A broader definition, provided at the Smart Cities Workshop (2009), defines a 'smart' city as '... a city that makes conscious effort to innovatively employ ICTs in support of a more inclusive, diverse and sustainable urban environment', a definition that is also adopted by the California Institute for Smart Communities (2001). An alternative approach for defining 'smart' communities' is to place placing emphasis on the importance of social and environmental capital in urban development. This implies communities, whose citizens are taught to learn, adapt and innovate. It has a strong focus on social inclusion and on participation in community affairs and decision-making processes in order to reach social and environmental objectives.

The terms 'community informatics' 'virtual community', digital community', and 'smart community network', used by various researchers, seem to have very close meanings to the CFG definition. 'Networked communities', on the other hand, relates to communities that have deployed digital broadband capability and make maximum use of it on behalf of their citizens, targeting economic development, organizational performance and high quality of living. 'Community networks' are defined as publicly controlled networks, at the service of the community, namely: individuals who use the network for communicating with friends, playing games, acquiring information, obtaining training etc.; and organizations running a variety of work tasks through the network. An 'intelligent community', on the other hand, is a community that perceives broadband communication services as a vital new collaborative opportunity for economic growth and social welfare. The distinctive attribute of such communities is the view of local users of ICT networks not as simple customers and consumers, but also as 'producers and creators of content, products and services'.

At least five different descriptions can be encountered for 'intelligent cities', as follows:

Intelligent cities are meant as virtual reconstructions of cities or virtual cities. The term has been broadly used as an equivalent to 'digital city', 'information city', 'wired city', 'telecity', 'knowledge-based city', 'electronic communities', 'electronic community spaces', 'flexicity', 'teletopia', 'cyberville', covering a wide range of electronic and digital applications relating to digital spaces of communities and cities.

- World Foundation for 'smart' communities defines 'intelligent' cities as 'smart' cities which, based on the adoption and use of ICTs, are paving a 'smart' development. This implies a conscious effort to use ICTs to transform life and work within a certain region (California Institute for Smart Communities 2001).

- 'Intelligent' cities were also defined as intelligent environments with embedded ICTs, targeting the creation of interactive spaces that bring computation into the physical world. From this perspective, 'intelligent' cities (or more generally 'intelligent' spaces) refer to physical environments in which ICTs and sensor systems disappear as they become embedded into physical objects and the surroundings in which we live, travel, and work.

- Along the same lines, 'intelligent' cities (communities, clusters, regions) were defined as multi-layer territorial systems of innovation that bring together knowledge-intensive activities, institutions for cooperation in learning and innovation, and digital spaces for communication and interaction, in order to maximize the problem-solving capability of the city. The distinctive characteristic, in this respect, is highly innovative performance, as innovation and solving new problems are the main features of intelligence .

The goal behind 'smart' city development is the provision of qualitative and innovative services to the public, to the economic activities, and also to the visitors of a city, together with the production of a safe, pleasant and inclusive urban environment. To this end, the development of a 'smart city' presupposes the proper *integration* of three layers:

- ⊙ *Physical layer*, incorporating human capabilities and knowledge-intensive activities;

- ⊙ *Institutional layer* that incorporates proper institutional mechanisms for social cooperation towards knowledge and innovation development. (More specifically it involves institutions and mechanisms for information diffusion, transfer of technology, cooperative new product development, etc.);

- ⊙ *Digital infrastructure layer* that incorporates a range of ICT infrastructure, tools, applications and content in support of both individual and collective action.

The concepts of 'smart' and 'intelligent' cities are treated in the available literature as being quite relative. However, in 'smart' cities, the emphasis is placed more on embedded systems, sensors and interactive media that support knowledge diffusion and interaction. 'Intelligent' cities, on the other hand, seem to rely more on collective /collaborative intelligence, innovation systems and web-based collaborative spaces. In both cases, the focus is on the integration of the three dimensions of urban space i.e. the physical, institutional and digital dimension.

## 7.5 GOING 'SMART': CRITICAL FACTORS

In its effort to explore the best practices amongst the world's Intelligent Communities adapting to the demands of the Broadband Economy, The Intelligent Community Forum (ICF), a think tank that studies the economic and social developments of 21st Century communities, defines five *critical success factors* for the creation of 'smart/intelligent communities', which are also used as evaluation criteria for assessing and rewarding the efforts undertaken by various cities towards 'going smart'. These are as follows:

- ⊙ Deployment of broadband communication infrastructure, used for the evaluation of the local capacity for digital communication. It should be noted that connectivity choices made by a 'smart' city need to be evaluated through both the prism of that city's local vision, and the affordability of costs incurred for the users.

- ⊙ Effective education and training of local labour force, strengthening high rates of adoption/use of ICT infrastructure. This results in increasing the capacity of this workforce to perform knowledge-intensive activities, while transforming 'individuals' into 'citizens', and enhancing the potential of participation in knowledge creation processes.

◉ Policies and programs that promote 'digital democracy' by bridging the digital divide among different groups of society, ensuring that everyone will reap the benefits of the broadband revolution (i.e. digital inclusion);

◉ Innovation capacity, assessing the level of creation of an innovation-friendly environment that attracts highly creative people and businesses.

◉ Marketing of 'smart' communities as advantageous places for living, working and running a business, which leverages the community's potential to attract talented employment and investments.

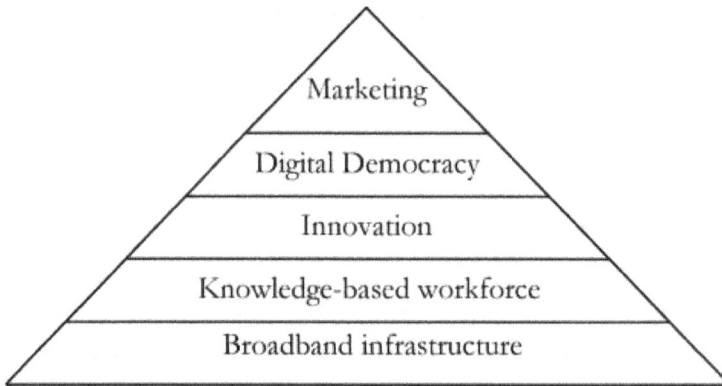

**Fig. 1: Critical success factors for cities 'going smart'**

*Source: Adapted from ICF website, also in Passerini and Wu 2008*

It should be noted that although technology forms the basis for community interaction, it is in fact the last factor that worries leaders of smart communities, as it constitutes a success factor that changes rather quickly. As experience from world pioneer smart cities shows, of critical importance are the rest of the factors presented in Fig. 1, as well as effective policies to challenge local people to use broadband. The key issue, in this respect, is not 'what' type of technology is available, but how 'effectively' this technology is used.

## 7.6 DIGITAL DIMENSIONS OF 'SMART CITIES'

The 'smart' city can offer local citizens and businesses a range of tools and ICT applications that can steer innovative behaviour. These applications create virtual environments, supporting both individual choices and group communication-collaboration options. The whole range of applications can be classified in the following groups:

◉ *e-Information*: refers to the provision of various types of information to a wide range of audiences, e.g. citizens, visitors, businesses, institutions;

◉ *e-Business*: refers to the potential offered to businesses for the exploitation of e-business opportunities, adoption of business-to-business (B2B) and business-to-client (B2C) interaction models, adoption of new innovative strategies for e-marketing their products etc.;

◉ *e-Marketing*: supports a range of e-marketing possibilities for a city / municipality in the promotion of the city's image (products, archaeological sites, cultural assets etc.);

◉ *e-Government*: refers to the provision, in a more effective way, of services to citizens, businesses, and governmental institutions (G2C, G2B and G2G interaction);

◉ *e-Innovation*: refers to the potential for e-cooperation and on-line development of new products;

◉ *e-Participation*: refers to the increasing potential for e-inclusion of citizens, thus strengthening active participation in the decision-making processes (e-Democracy).

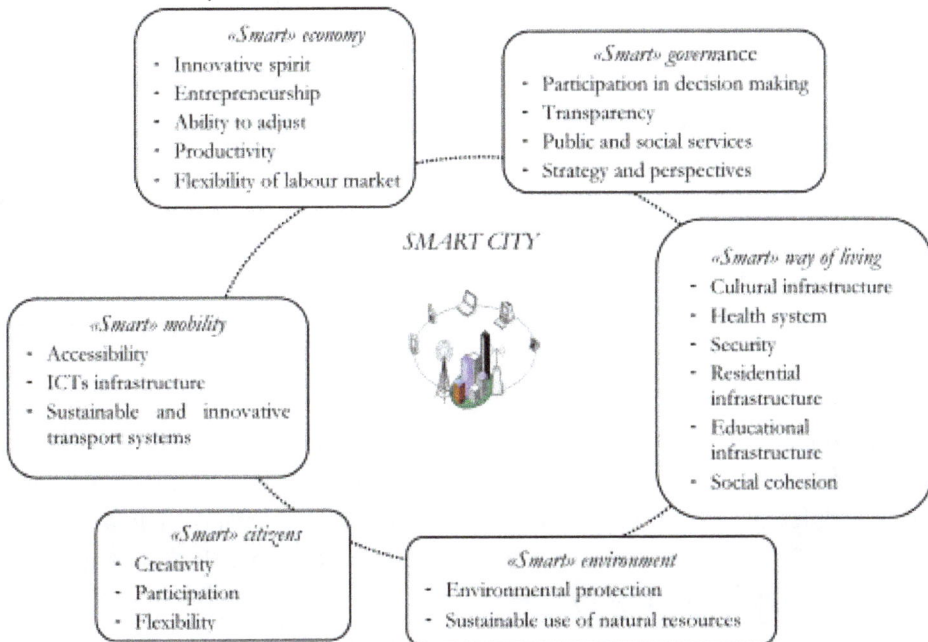

**«Smart» economy**
- Innovative spirit
- Entrepreneurship
- Ability to adjust
- Productivity
- Flexibility of labour market

**«Smart» governance**
- Participation in decision making
- Transparency
- Public and social services
- Strategy and perspectives

**SMART CITY**

**«Smart» mobility**
- Accessibility
- ICTs infrastructure
- Sustainable and innovative transport systems

**«Smart» way of living**
- Cultural infrastructure
- Health system
- Security
- Residential infrastructure
- Educational infrastructure
- Social cohesion

**«Smart» citizens**
- Creativity
- Participation
- Flexibility

**«Smart» environment**
- Environmental protection
- Sustainable use of natural resources

**Fig. 2: Dimensions of smart city development**

*Source: Adapted by Tsarchopoulos (2006)*

Based on a search of 'smart cities' literature, it is evident that this term does not carry a holistic meaning gained by the integration of certain city attributes/functions. Instead, it is used to describe innovative aspects available to a city that are based on the adoption/ use of ICTs. These aspects can be associated with the economy, local population, governance, citizens' participation etc. With regard to the economy, a 'smart' city can be a city hosting a 'smart' industry (i.e. an industry that is either a producer or a heavy user of innovative ICTs,) or a city which develops highly ICT-based business parks in its territory. It is also used to describe a city with 'smart' inhabitants (i.e. highly educated local human resources, or a city where G2C (government to citizen) interaction is heavily based on ICTs (e-governance); or a city exhibiting a strong, ICT-enabled, public participation in local decision-making processes (e-democracy). Moreover, it may refer to a city which makes use of modern ICTs in urban processes in order to improve the quality of life for its inhabitants (e.g. 'smart' transport systems in support of urban traffic management). Finally, the term is used to describe a city that makes use of ICTs to improve services in several fields, e.g. security/safety, health, 'green' urban development, or sustainable energy consumption.

## 7.7 SMART CITIES' AND COMMUNITY DEVELOPMENT

Getting access to effective and affordable ICT systems is crucial for reaping the benefits of communication and driving *community development processes* in the broadband economy context. This places new challenges in front of planners and regional policy-makers regarding the bridging of the *'digital divide'* at community level, and coping with *ICT illiteracy*. The UNESCO Report (1980) entitled 'Many voices one world', stresses the need for a *'democratization of information flow'*, implying more equal access to information for larger groups of society, together with the need for policy action towards this end, dealing with the development of high-quality broadband connectivity. It also articulates and legitimizes the idea of *'the basic human right to communicate'* and be informed about whatever might affect daily life in order to support autonomous decision-making. Moreover, emphasis is placed on the role of citizens and stakeholder groups as *carriers of change* at the community level, based on participation and access to information.

Today the scope of citizens' and stakeholders' right to information has been considerably broadened, in alignment with the huge development of ICTs and their applications that has given rise to a further increase in interaction and networking potential both between and among different groups of society, thus contributing to a more *equitable share* of knowledge and information, and a relative shift of *power relationships* affecting decision-making processes at the community level.

A crucial *objective* of the 'smart' cities' perspective relates to *community development*, seeking to empower local individuals and groups by providing them with the necessary skills and information to affect changes in their own communities. As Lee (1989) states, community development represents a *process of change*, where *participation and collective action* is of crucial importance. It implies a community where individuals are assisted to acquire skills and competences and to develop their own views and attitudes, a requirement for their democratic participation in a wide range of community problems. Community development creates the foundation for building communities that are based on justice, equality, mutual respect and cooperation; it also forms the cornerstone for the creation of *relationships and networking*, thus strengthening bonds and understanding among local citizens, as well as the creative exploitation of local knowledge and experience. Moreover, it is considered as a step for influencing power relationships and their role on policy decisions, by changing the position of ordinary people and their potential to affect local decision making. The key aspects of community development, according to Lee (1989), are presented in Figure 3, where the role that 'smart cities' can play towards this end is also shown.

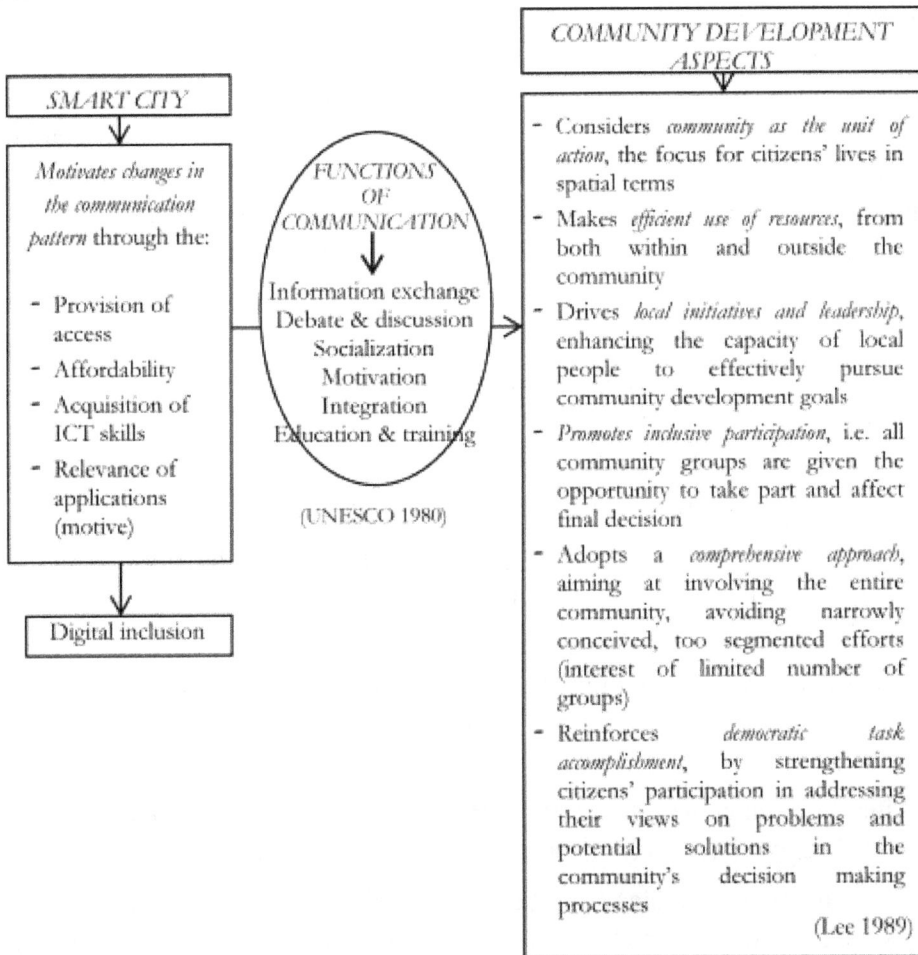

**Fig. 3: Smart cities for community development**

In the last two decades, the concept of "smart city" has become increasingly popular in international scientific and policy literature. Anthopoulos (2017) defined that smart city is an urban space that managed and operated by smart systems as well as inhabited by clever insight/ideas of citizens. ICT (The Digital or Information City) and its innovations are utilized as a means to achieve the sustainability of economic, social and environmental aspects referred to solving the problems on the dimensions of people, economy, governance, mobility, environment and living. Smart Cities initiatives aim to "provide more efficient services to citizens, to monitor and optimize existing infrastructure, to increase collaboration amongst different economic actors and to encourage innovative business models in both private and public sectors".

## 7.8 SUPPORTING FACTORS AND THE DIMENSIONS OF SMART CITY IN THE CONTEXT OF URBAN PLANNING

As stated that there are eight factors related to the interrelationship between urban development and Smart City. The eight factors consist of: community, government, economy, technology, organization, policy, natural environment and infrastructure development. Three out of eight factors are considered as the main factors in managing smart urban planning i.e., Technology, Organization and Policy. The three mentioned factors are interconnected with each of the other five factors i.e., Governance, People communities, Economy, Natural Environment and Built Infrastructure. Scientists emphasizing on some aspects that should be supported by smart technology. The aspects are as follows:

- Mobility dimension (smart mobility): in which the city is quipped by smart parking, traffic light, bike, driverless bus/car, electric and hybrid cars, and active transport.

- Services dimension (smart urban service): in which the city is facilitated by the systems of smart wastes, lights, parks, and gardens.

- **Government dimension (smart government): in which the city is governed by** smart system of administration, payment, data sharing and business services.

- Inhabitant dimension (smart citizens): in which the citizen is facilitated by the smart system of community services, tourism and major events, civic app (social networks, NBN and public WIFI), digital hubs, libraries, citizen involvement, and labs.

- Building component dimensions (smart building): in which the city consists of intelligence buildings completed by smart infrastructure.

- Environment dimension (smart environment): in which the city consists of smart system of environmental monitoring, electrical cars and charging stations, as well as renewable energy.

- Public health and safety dimensions (smart public health and safety): in which the city has intelligence systems in managing incident, security services, health and human services (smart aged car, smart homeless reduction)

- City development planning dimension (smart city development planning): in which the city is managed as smart growth and public realm.

- The smartness of a city is related to the components that must be measured from each factor below:

- ◉ Smart economy is measured by the level and variety of competitiveness (innovation spirit, entrepreneurship, productivity, flexibility of labor market, international contributions, ability to transform);

- ◉ Smart people are measured by the level and variety of social and human capital (level of qualification, affinity to lifelong learning, social and ethnic plurality, flexibility, creativity, cosmopolitanism/open-mindedness, participation in public life.

- ◉ Smart governance is measured by the level of participation (participation in decision making, public and social service, transparent governance, political strategies and perspectives)

- ◉ Smart mobility is determined by the service level of transport and ICT (local and international accessibility, availability of ICT-infrastructure, sustainable, innovative and safe transport system);

- ◉ Smart environment is concerned with the sustainability of natural resources (attractiveness of natural condition, pollution, environmental protection, sustainable resource management);

- ◉ Smart living is measured by the quality level of life (cultural facilities, health conditions, individual safety, housing quality, educational facilities, touristic attractiveness and social congestion).

As highlighted four dimensions that should be included in the urban planning element of smart city-based approach. Each of the four dimensions is sustainability (covers the problem solution of infrastructure & governance; energy & climate change; pollution & waste; social, economic & health), quality of life (covers the human guarantee of financial well-being, emotional well-being), urbanization (covers the term of technical, infrastructure, governance and economics), smartness (covers the aspects of economic, social and environment).

# 8

# URBAN PLANNING MANAGEMENT

## 8.1 INTRODUCTION: AN OVERVIEW

Nowadays, we see increasing and creating new areas to be small cities or suburban around the big cities, so the question is how to balance between different original areas services, "Urban-rural integrated planning is to regard economic and social development in urban and rural areas as a whole for uniform planning from the integrated situation of national economic and social development and make integrated research and integrated solution to the existing problems and their interrelationship in urban and rural areas so as to change the urban-rural dual structure gradually and establish an equal and harmonious urban-rural relationship under marketing economy. Also mentioned although there is associativity between the theory and practice of planning, both of them are suffering of being far away than each other. More precisely, some of the plans still discuss outside the central authorities. Thus, many factors become more weakness than others "Consequently, strategic planning has diffused widely in this industry over the last decade, and is an important component of hospital management, As example of being the management is the cure that helps to manage the plans such as transportation management planning which is considering backbone in the cities or all communities. As saw that "It was an important to improve urban transportation management level by carrying out urban transportation management planning. According to the characteristics of urban transportation management planning, the evaluation model of urban transportation management planning project was studied based on fuzzy theory.

*Urban planning considers one of science that gathers set of the fields in one point. Management, Art, geography, ideology, psychology, Topography and more are collected in the same side because dealing with the environment that contains human being, animals, land, weather and anything can be seen in this world is very complicated, and it requires many attempts and huge data to get a typical solution.*

First, the 14 evaluation index of urban transportation management planning project was constructed from three aspects, which are safety, smooth and environment.

To approve that there are many regions that were interesting to be one of organized regions as what mentioned in this article, "Urban management has become a key that is required by Chinese information technology strategy while digital urban construction is an important means to promote urban management,". Information technology strategy while digital urban construction is an important means to promote urban management."

Furthermore, to support what are said about the management in this aspect, many projects could be faced by many dangerous if there is no exact managing to this area, "Urban storm water, as an important water resource, has aroused the concern of many experts at home and abroad. Environmental ecology, urban planning, landscape design, engineering and other issues have been involved in the management and development of urban storm water,."

In fact, more definitions have been made to treat the quality management to be very close of planning with knowing what should be done. As referred that "Integrating quality management practices with organizational knowledge concepts can provide insights into how quality management lead to improved performance. Moreover, many of decision makers do not know the connection between using planning management and the benefits that should be gotten from it, Chen discussed that "Changing land use and land management practices are therefore regarded as one of the main factors in altering the hydrological system, causing changes in runoff."

Furthermore, the world and especially the planners will be in critical situation as be referred in this article that written be Campball "In the coming years planners face tough decisions about where they stand on protecting the green city, promoting the economically growing city, and advocating social justice. Conflicts among these goals are not superficial ones arising simply from personal preferences. Nor are they merely conceptual, among the abstract notions of ecological, economic, and political logic, nor a temporary problem caused by the untimely confluence of environmental awareness and economic recession."

Also, to explain many examples that clarify what does mean by saying a risk as mentioned by Gheorghe "Thus, fire risks prevention and reduction represent an integrated assembly of specific activities of management, evaluation and control at state level based on a risk assessment and covering plan included in the national firefighting strategy."

"Environmental planning and management of peri-urban areas is informed by three distinctive fields, namely rural, regional and urban planning, and the multitude of traditions that characterize the evolution of each.". Then, treating what has been done could take more time, sources and tools that also should cost the countries to fix it or solve it, David said in this "Comparison of historical: and current aerial photos reveals changing land use/land cover patterns, which may also assist in making appropriate urban planning and management decisions." Also, Yewlett referred that "Most of the 'Strategic Planning' with which OR practitioners are involved concerns industrial or commercial strategic planning; essentially, the planning by an organization of its own intended strategy,".

Fallowing managed ways is very important to increase the productivity in the whole system of knowing the relationships between planning a city and knowing how to lead it and how to be working better. Gold said important words in this side " Productivity measures have fluctuated between those that are overly fragmented and those that are overly aggregated. Both fail to meet management's need to understand the links between partial effects and combined effects in order to identify the causes of improvements as well as the factors limiting their benefits."

Many styles are still satisfied today by a lot of governmental and private organizations, but to be worked together is still uncommon because there is no mechanism to lead that, as Brody discussed " Co-system management represents a departure from traditional management approaches by addressing the inter action between biotic and abiotic components within a land or seascape, while at the same time incorporating human concerns."

## 8.2 BUILD A MODEL WITH FACTORS

The model shows the connection between four major that to be the circle around studying urban planning and management.

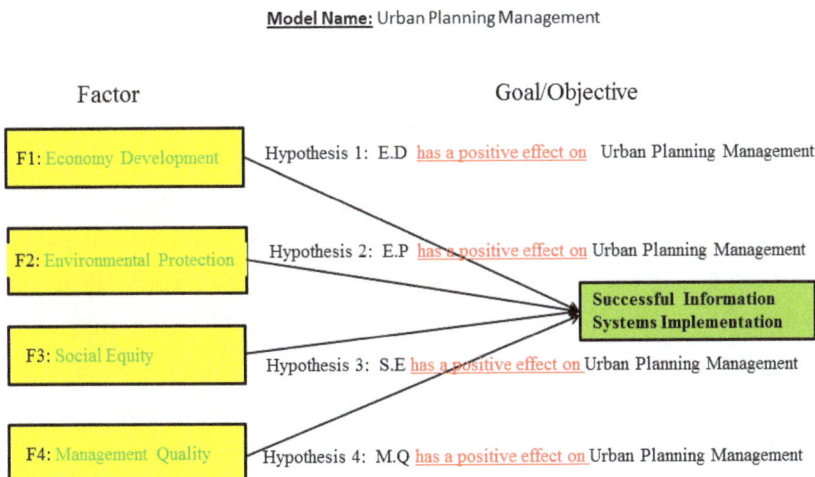

**Model Name:** Urban Planning Management

**Fig. 1: Shows the relation between factors and objects or goals.**

## Quotes from References

### Factor No (1): Economy Development

"This development conflict also happens at the local level, as in resource-dependent communities, which commonly find themselves at the bottom of the economy's hierarchy of labor. Miners, lumberjacks, and mill workers see a grim link between environmental preservation and poverty, and commonly mistrust environmentalists as elitists."

The author persuasively argued that economy development is one of the important levels that affects deeply in how people live in their communities, and life that they will live it with economic changes. Therefore the factor is important to consider as factor that control in several sub factors that also be as means to build our communities. Furthermore, to be considered in our world that is defined as cities and population, we have to be aware that while the world gains more people every day, there are many needs that must be found for them.

### Factor No (2): Environmental Protection

"The Major of Urban and Rural Planning and Resources and Environment Management (following referred to as MURPREM) is a relatively new interdisciplinary subject which, on the basis of the geographical science, environmental science, planning sciences and management science, was set up by the Ministry of Education of P.R.C. in 1999."

The author and others strongly suggested that environmental protection definitely had an impact on the success of managing our plans because our land is the one source that we can get benefit of it, and also it will be the solution to many of problems that could face the world today or might be tomorrow.

### Factor No (3): Social Equity

"In the social sphere, various social movements (*women's liberation, the environmental movement, the student revolt, etc*) led to a startling shift in values and beliefs. Major aspects of the traditional Swedish devotion to progress were called into question. Planning under increasing uncertainty and decreasing resources resulted in conflicts between agencies, which became less and less willing to implement solutions which would ultimately reduce their domain as well as status. There are several implications of this development."

Here, it has been said in several articles about sociality and how to be fair to find rules and constitutions that help people to equal in their land, but this factor is very hard to be understood because it requires many efforts to reach to this equality.

### Factor No (4): Management Quality

"Artistry is an important aspect of reflecting the characteristics of the storm water management system. Every city should create characteristic system according to the city's cultural background, ecological background and construction background. Storm water management technology should be combined with landscape design. The government can promote the urban landscape quality based on landscape design and storm water management."

The author mentioned in this side with others about one of the important factors that the whole countries and systems around the world still find it complicated and requires more hard work and rules to be existed. Despite of studying and making a lot of the rules in this science, but it is still here and there many of corruption that face our organization and countries.

## 8.3 THE EVALUATION OF INFORMATION SYSTEM: ISSUES AND CONCERNS

As wherever either in the rich and big countries or poor and small others dealing with urban planning is one of the concerns that is facing the communities today. According to many of the reports and surveys that have been done in this field, there is very important to be aware the relationship between urban planning and management. Moreover, several planners recently tried to link all of the factors with urban planning. "While much research has been geared toward instituting the broad principles managing natural systems, comparatively little work has been done to evaluate the specific tools and strategies involved in implementing ecosystem management.

Also there are specific limitations that consider the differences between a lot of communities and countries, for example as Alcorn referred owning or working on some of the powerful energy as nuclear energy or wind turbine energy gives the owners more flexibility to be on the top of an organized countries if it has been known how to occupy these chances. Furthermore, to start with limited equipment is better than standing at same situation, "Land use of arrangements and investments for the people to take some land improvement, production, maintenance or change it, so the marriage between urban planning and management is absolutely the solution to many of missies the have been seen in most of cities and states, "Urban planning management plays an important role in effective integration, standardized administration and authorized share of planning data, which could effectively resolve universal problems such as nonstandard format, resources dispersion, etc.. To prove what should be done, there is factors that help in growing these societies whatever is people, land use, transportation and others, as result this need could be understood by specialists but other might not. "Urban planning management is the top-level work of modem city management.

## 8.4 EVALUATING A MANUFACTURING INFORMATION SYSTEM. CONCEPTUALIZATION

To be evaluated, a system that should be able to figure out the difficulties that turning around some cities and villages has to compare between the fundamentals that any region stands on them, and the risk the has been face if did not do it. Stone mentioned that

"When there are multiple scenarios, one must assign probabilities or credence's to each scenario. Since each scenario corresponds to a probability distribution, the resulting target location distribution is a mixture, weighted by the scenario probabilities, of the scenarios. In spite of urban planning is multiple majors, using the management in creating advertisements that will be helpful in recognizing the overall bases. Also, treating all the problems inside the cities could be as transportation, population, infrastructure and more, so it will be as be said in that by Ma, "Urban transportation management planning is used to analyze and assess all objections of transportation management from aspects of administrative and technical management. Next, planners know that to begin developing area or city, they have started to improve the human being how live in these areas. As see that "Urbanization, a development of human society, is the inevitable result of rapid emergence worldwide.

## 8.5 FINANCIAL JUSTIFICATION

As known in Architecture schools, there is rules and conditions that give each city more privacy. This privacy consists of the density of people how live there and the level of their live. Newman and Hogan referred that "The trend to lower urban density is fueled by the belief that high-density living is unhealthy and stressful. Thus, making a benefit of using some of material that could be used in an area is one of the financial supports that assists the cities to be able to grow up normally.

For example, Li and Jin said that "At present, the employment situation is very austere in cities of China, and we need to exploit a new way about employment. At the same time, there are 150 millions of residual labor forces from countryside need to transfer ." That means also labor power is very important to plan and manage a cite to become in future a place has ability to receive people. "Even though, the planning and management departments were aware of standardizing the examination and approval of outdoor advertising, depending on public officers' strict management is not enough. To approve what has been seen in this field of building cities on high level of modernity there are some of instructions that require to work on and gather them in one way to figure out the solutions that should be taken in this order, where others said that modern urban infrastructure needs a comprehensive and integrated information science and technology.

Furthermore, one of the reasons that make planners working on transportation as a major part of whole project that goes back to the significant this part instead of others, for example rush hour is difficult happen that planners and governors are willing to solve these wasted time and energy in most of big and middle cities, so it considers as central point must be taking care of it. According to many studies, the numbers of vehicles that are moving on the roads increase sharply every year. The new driver rapidly growth is another major cause of traffic accidents frequency .The annual growth drivers is about five million. According to statistics, more than 50% of the traffic accidents is the responsibilities of drivers with less than 5 driving years.

Based on what have been said about the majority of using the roads and how transportation is effect in the style and economic of each country. As says in that "The transportation is the national economy basic industry, no matter the past, the present or the future, the transportation industry will be the human society progressive important material base."

## 8.6 LIFE CYCLE EVALUATION

Although urban planning with its subset are being refreshed every year, it also has time cycle which is depending on the nature of the area that should be planned on, as example of one part of what is discussed mentioned by Williams and others "Species of most life forms had a consistently higher probability of local extinction at urban grasslands than at grasslands in per urban landscapes, and rural landscapes had the lowest probability (between 0.1 and 0.2 for all life forms)."

Also, the life cycle evaluation assists in knowing what have been created and how to select factors that as be mentioned is extracted of the area where is worked on, so W.J and others see that "The collective construction land in the urban planning area is an important component of urban land. According to incompletely statistics, its number accounts for about 50 percent of urban planning area's land."

On the other word, to approve collecting data of what turning around these cities, there must be system or mechanism to introduce the variables that classified as important points in different areas, Wang said that "Urban Planning and Management Information System (UPMIS) comes from west nation, took Maim Frame as hardware terrace in the 1970's at the earliest stage, asked for help of transact information system software, the establishment database in the city." In general, Urban planning is very complicated system, and it requires many sections to work in high level of efficiency, Jinxing and others agree with this point, and they said that "Urban system is an enormous and complicated system; various types of activities carried out in the urban system have complex relationship with the population travel. Through GIS, traffic spatial analysis can be done by traffic survey with social attributes, spatial attributes, and spatial behavior.

## 8.7 RESEARCH METHODOLOGY

This paper tried to discuss many of ideas and values that could be lost in the past. "Value Sensitive Design is a theoretically grounded approach to technology design that accounts for human values, such as privacy, fairness, and democracy, throughout the design process. The method has three key features: an interactional perspective, attention to indirect as well as direct stakeholders, and a tripartite methodology." Also, it is necessary to see the urban planning as a policy decision; especially, when the process requires some procedures that must be done by politicians, so it has said in this case by Jiao that "The democracy in planning has been greatly promoted.

In order, discovering the sources that feed any of places that human being live in, it is supposed to take the water as main source, so it must be at the top of the list when studying any project. Zhang and others referred to that "Urban storm water, as an important water resource, has aroused the concern of many experts at home and abroad."

Actually, environmental science, planning sciences, geographical science and management science are the main aspects in urban planning, so all these fields based on each other, and urban planning covers all of them.

After that, it can be said there are many links that connect between urban planning and management, all of them reflect on people and computerized system, where overall the data that must be collected in several ways, it should be by people and the used system. On the other hand, dealing with many sources that one of them is transportation could be more affective when the planners know how to take it as important point, Cannon mentioned that "Only when we learn to understand much better the dynamic interaction between new transportation systems and the structure of our communities can we take full advantage of transportation's potential as a really effective tool in our quest for a finer quality of life for our citizens.

## 8.8 RESEARCH FINDING

This paper has reached to accepted level even it has been gotten on results that is not expected at first. The results are to move from point to another because this case is not a simple study but it is a complicated one. Many studies as this were in different regions, as sees that "Across the conterminous United States, the WUI covers 719 156 km² (9.4% of the land area) and contains 44 348 628 housing units (38.5% of all housing units). All 48 states contain WUI areas, but the eastern United States has the greatest extent, especially in northern Florida, the southern Appalachians, and coastal areas of the Northeast.

To complete the preceding one, the paper supports several points that are explained in many studies, for example Don sees that "Greenways in urban areas have the potential to provide a unique combination of ecological and social benefits to the metropolitan region. Ecological benefits may include stream quality and wetland protection, erosion and flood protection, habitat and gene pools of native flora and fauna, and air quality and microclimatic improvement."

Also, this study passed through how to measure some factors which are very important to make any project useful for what has been done to, it is remote sensing technology that has great role in urban planning. Next, the paper considers the rate of traffic that moves on the roads in different regions depends on the communities and how is built these cities.

## 8.9 CASE STUDY ANALYSIS

In fact, this linked in one side how to improve decision making about rural lands that requires careful consideration. Moreover, well-known information or how to be affected with the areas wherever is especially when the planners are making important decision that costs a lot of money. Last studies discussed like these thoughts, so Jon said that "The work is timely because land use planners need tools to assess communities."Despite unprecedented federal investment" natural science planners who received their professional educations in the late 1960s and early 1970s deal ineffectively with "public participation" required by the Clean Water Act and Coastal Zone Management programs."

Furthermore, the paper assumes that many of treatments could be given when the system is facing some difficulties, but the best way here is building the system on right foundation. Also, this paper referred to use some of technologies that have been tried in this field, and it can be useful and efficiency with all the risks that might be happened in variation time. As in this said that "GIS has been widely used in solving a variety of planning problems. In the process of transportation planning, omnidirectional transportation information collection, investigation of city transportation situation is foundation and prerequisite of realizing the scientific transportation planning."

# 9

# INTEGRATIVE THEORY APPROACH TO SUSTAINABLE URBAN DESIGN: THE VALUE OF GEODESIGN

## 9.1 INTRODUCTION

Sustainable urban design connotes a new relationship between the natural environment, urban form and structure, economic and institutional processes, and social livelihood. It requires a transformation of the existing socio-economic, environmental and urban design settings. Atkinson and Ting (2002) conceptualize sustainable urban design as an attempt to recognize the complex and hitherto-neglected relationship between the natural environment (sustainable) and the city as artefact (urban design). It seeks to enable the natural processes that sustain life to remain intact and to continue functioning alongside initiatives for the improvement

*The concept of sustainable urban design is an overarching concept that can serve as a platform to resolve the conflicting values of the traditional urban form and modern design models. However, the principles of these models both traditional and modern should be integrated with the sustainable urban design principles to effectively incorporate them in urban planning and development. This chapter has tried to highlight some of the pertinent and core principles of traditional urban form and sustainable urban design that should be integrated to foster liveable cities. It also highlights the importance and value of GeoDesign to sustainable urban design.*

of individual quality of life and the well being of the society. Sustainable urban design adopts a systemic and synergistic reorganization of environmental, economic and socio-economic goals that enhances the long-term health of natural systems and the vitality of urban communities. The concept of sustainable urban design requires a comprehensive framework of new urban design ethic to promote sustainable cities. Different authors have elaborated on the frameworks and guidelines of incorporating the principles of sustainable development in urban design. However, there is no agreed strait-jacket framework of sustainable urban design. The context in which the principle is applied determines the form of sustainable urban design. The challenge is to develop the appropriate urban design guidelines for a particular local context.

In metro cities, the spate of modernization has led to the replacement of traditional urban structure and form by Western models of urban form and design. This has resulted in problematic urban development as Western models are adopted without recourse to the underlying principles and socio-cultural background of the traditional form. In the drive towards sustainable cities through design, the challenge is to develop a framework that will adapt traditional urban form to changes in the face of Western models of urban form. This chapter proposes that there is a need to reorient the approach to urban design and development in favour of an approach that is integrative in terms of theory and provides for sustainable development. It examines the urban design problems in Saudi urban development, highlights the sustainability issues, proposes an integrative framework to address the issues and lays out the basic parameters of the framework and some cases of its application.

## 9.2 THE INTEGRATIVE THEORY APPROACH

The integrative theory approach was suggested by Sternberg (2000) in an effort to establish a theoretical foundation for urban design. Sternberg (2000) observed that urban design had been relying on techniques and ideas that have no clear theoretical basis and suggested an integrative approach to defining the foundations of urban design. He posited that "ideas that inform urban design usually coalesce around contending approaches" and shared principles of these approaches should be integrated to establish a general theory of urban design. Sternberg (2000) highlighted four elements of integrative urban design that include good form, legibility, vitality and meaning. The principles are mainly related to the substantive aspect of urban design due to the need for "a complement to procedural theory: a substantive planning theory that sheds light on the specific concerns of the urban designer" .

Sternberg (2000) highlighted some criteria (referred to as challenges) that an integrative theory should fulfill. The set of criteria include highlighting the underlying principles of contending approaches, addressing substantive urban design issues, awareness of the "constituents of human experience of built form", unifying economic and architectural traditions and being realistic and practical. These criteria are used as reference in developing an integrative framework for sustainable urban design. Urban design principles in Saudi could be considered to be generating from, at least, three sources; traditional urban design principles, contemporary or conventional urban design principles and recently emerging sustainable urban design principles. The idea is to integrate the underlying principles of these sources

## 9.3 SUSTAINABLE URBAN DESIGN: A PARADIGM SHIFT

### 9.3.1 The rationale for sustainable urban design

The spatial organization of cities in terms of structure and forms is rapidly being influenced by economic forces at the detriment of social and environmental factors. For this reason, cities are characterized by physical and environmental problems in terms of inadequate infrastructure, deteriorating environmental quality and congestion. Urban problems do not arise from the inherent nature of the cities but due to the absence of effective urban governance and management (Jenks and Burgess, 2000). Planning and urban design are interlinked with the dynamics of urban transformation and have been recognized as having a vital role in the management of urban development. Land use planning and urban design influence urban structure and form which eventually generate social and economic activities within the city. The BEQUEST framework for sustainable urban development identified urban design as one of the activities that influence sustainability. It also observed that the manipulation of urban form, and the provision of better forms of governance, may go some way to overcome city problems. In addition, the study by Banister et al. (1997) concluded that significant relationship exists between energy use in transport and physical characteristics of the city such as density, size and amount of open space.

Further empirical studies have shown that urban form and structure influence the social configuration, economic efficiency and environmental performance of the city. It asserted that urban configuration influences outdoor climate conditions, energy balance of building and diffusion of pollutants while Burton (2000) highlighted the negative and positive influences of urban compactness on social equity. The findings by Cervero (2001) suggested that the urban form and mobility characteristics of metropolitan areas have some bearing on economic performance. In essence, there is an interrelationship between the spatial, physical, and structural characteristics of a city and its functional, socio-economic and environmental qualities and this relationship should be explored to foster liveable cities.

Traditionally, urban design considers the relationship between urban structural elements, socio-economic activities and environmental quality. As elaborated on the visual quality of the city and highlighted elements that are crucial to the image ability of a city. His emphasis on the interrelationship of these elements and the physical environmental quality of the city became analytical means of promoting city live ability. It stated that the physical design of the city cannot be isolated from the problematic context of existing cities and highlighted the problems that concern designers as economic, engineering considerations, social and ecological. Evidently, the emergence of the concept of sustainable development has boosted the incorporation of social, economic and environmental dimensions in urban design process. The principles of sustainable development require a balance consideration of social, economic and environmental implications of development activities. Urban designers seek to incorporate the principles of sustainability into urban design through sustainable urban design.

In other Middle East countries, there is growing awareness of the unsustainable water and energy consumption. Domestic energy demand is increasing due to automobile dependence and use of energy dependent air conditioners for cooling. Saudi Arabia consumes about one third of its oil production and buildings are consuming about 30% of domestic usage. Water production by desalination depends on fossil energy and high water demand influences energy demand. There are indications that the usage of water has not been efficient due to wastage and high energy consumption and automobile dependence might have led to air and noise pollution in a typical Saudi city. Therefore, there is the imperative to charter a new course that is more sustainable. As highlighted the recent drive by Middle East Governments To Start Initiatives That Will Foster Sustainable Built Environment.

## 9.3.2. Pattern of Urban Development

### 9.3.2.1. Traditional urban form

The traditional urban form in Saudi Arabia is similar to that of most traditional Muslim cities. The traditional urban fabric is characterized by organic narrow winding street pattern, with homogeneous arrangement of housing plots. The houses open inward in form of courtyards (Fig. 1) and are centred on mosques, markets 'suqs' and madrassas. As noted "the formation of the urban structure is not subject to the purely quantitative division of large space into smaller fragments but based on an incremental or 'organic' aggregation process, originating in the definition of socially relevant micro-spaces which are then connected into larger units. The enclosure of voids by correlated solids, repeated in countless variations, is the generating principle of urban form". Highlighted the elements of traditional Islamic city as the obviation of the need for public buildings; the centring of city on mosques that provide a range of welfare and education functions; the bazaar or 'suq'; the residential fabric that is composed of a compact structure of open courtyard houses; and the irregular street pattern. The irregularity of forms of the traditional urban fabric does not necessarily connote lack of order but depicts coherent and harmonious integration of diverse elements to make a whole.

**Fig. 1: Traditional courtyards**

In spite of the general elements attributed to the traditional urban fabric, there are notable variations from place to place. The old Jeddah as a typical example of traditional town in Saudi Arabia does have some specific characteristics that are peculiar to it. The old Jeddah town was significantly different from other Islamic cities by its lack of central space allocated to governmental or religious institutions (Fig. 2). The core of old Jeddah emerged around the central 'suq' or market surrounded by residential quarters (Khan, 1981). Nevertheless, the social and communal activity still centred on the mosque. The main arteries are very few numbering about five, including the major axis along the 'suq' or market. The width of the roads varied according to function and location. The narrower lanes were located within residential quarter (Fig. 3) while the wider streets served the shopping areas and transportation of goods.

**Fig. 2: Street patterns in old Jeddah (Source: Google earth)**

**Fig. 3: A narrow street (buildings have traditional wooden window screen "Mashrabiyah" for privacy)**

Religious, environmental, socio-economic and cultural factors have been cited as having influences on the elements of the traditional urban fabric. For instance, the introverted housing pattern of courtyards have stemmed from the religious concept of privacy and adaptation to the local climate. Also, the irregular street pattern reflects an adaptation to the local climate by maximizing shade. In essence the traditional urban fabric exemplified adaptation to local environment, integration of socio-economic and religious-cultural principles in developing harmonious and liveable society. The balance of socio-economic, environmental, religious and cultural factors in development of traditional urban fabric is exemplary and noteworthy. In the first instance, the origin of the city could be based on environmental, socio-economic or religious considerations. Availability of water or good agricultural land could serve as considerable environmental factor for locating traditional cities. After which the city is developed in an incremental manner without a 'formalized' planning but with a general concept of harmony, coherence and liveability.

The spatial geometry of the traditional urban fabric seems to have developed from lack of planning. Far from that, the structures are planned but the planning principles are flexible enough to allow for acceptable diversity and the principles are applied by the individuals in the society as there was limited civic planning. The main sources of these principles are the religious tenets derived from the Shari'ah. Examples are the principles of privacy, private and public space. The principle of privacy might have contributed to the development of the narrow and winding streets apart from the climatic adaptation by "shading".

### 9.3.2.2. Contemporary Urban Form

The emergence of contemporary urban form in Saudi Arabia started in the 1930's when building regulations were enacted to guide building construction and street patterns. During this period, imported modern technologies and planning models were introduced to the country without due consideration of the local traditions and socio-cultural factors. Notable among the contemporary building regulations were the 1358/1938 King Abdulaziz's order to found Alkhobar city, the 1371/1951 ARAMCO home ownership plan and the 1960 circular by the Deputy Ministry of Interior for Municipalities. These orders and regulations set the background for the contemporary urban fabric in Saudi Arabia and the structure and pattern of cities and towns are influenced by the different regulations. Greater degree of urban transformation set in during the 1970's as a result of the economic boom and the inauguration of the Five Year Development Planning. Then, the government began a campaign of modern urban planning and systematic intervention in urban production. The new urban form was established with the grid-iron patterns and building regulations and zoning outlined compulsory setbacks and site-coverage limits. The new spatial models engendered the construction of freestanding, low-density "villa" dwellings .

Al-Malaz neighbourhood in Riyadh represents a typical Saudi contemporary urban structure. The neighbourhood, which is located 4.5 km north east of Riaydh, was planned in 1373/1957 when government headquarters was moved from Makkah to Riyadh. Al-Malaz was planned following a grid-iron pattern with an hierarchy of streets, rectangular blocks, and large lots which in most cases are square in shape (Fig. 4) (Al-Said, 1992). The main thoroughfares are 30 meters in width, secondary streets 20 meters, and minor streets or access streets of 10 and 15 meters. The block areas are 100 by 50 meters. The typical lot size is 25 by 25 meters, with some variations in width. The Al-Malaz neighbourhood structural pattern was consequent upon the contemporary building requirements which stipulated the planning of the land, subdivision with cement poles, heights of the buildings, setbacks and square lot ratio of the buildings.

**Fig. 4: Structural pattern of Al-Malaz neighbourhood (Source: Google earth)**

### 9.3.2.3. Issues and problems in contemporary urban form

The contemporary urban pattern is mainly driven by economic considerations and formalized planning legislations. The streets are widened (Fig. 5) to maintain fast connectivity among different sectors of the city through the automobile. Urban development activities are evaluated mainly by economic efficiency and traffic considerations with the neglect of socio-cultural and environmental dimensions. The contemporary model of urban design encourages the extensive use of space and the fragmentation of functional spaces. In essence, the contemporary model contrasts the traditional model by being dynamic and mechanical while the traditional model is static and human in scale.

The contemporary/modern model of urban form has been found to be in conflict with some indigenous socio-cultural, environmental, economic and structural concepts. For instance, in the traditional Arab-Islamic society privacy was very important but the introduction of setbacks allowed adjoining buildings to open their windows outward thereby infringing on the privacy of other dwellings. Also, the introduction of glazed glasses as building materials results in additional costs of cooling and heating during the extremes of climate in summer and winter. These notable conflicts rendered contemporary model of urban design to be problematic. The residents have rejected contemporary urban form by erecting additional structures over fences to ensure privacy and by not using their yards for female activities. Social sustainability is also affected by the use of cars for movements within neighbourhood such as going to school, mosque or shops instead of walking. Thus, it opined that the new mix of Western styles of design and characters that have appeared recently have changed the spatial environment of many Islamic countries for worse. The challenge is to develop a framework of adopting the modern technology and design principles without jeopardizing the elements of traditional values, forms and design.

**Fig. 5:A Wide Street of the contemporary urban form**

# 9.4. OVERVIEW OF SUSTAINABLE URBAN DESIGN PRINCIPLES

It is recognized that urban design could foster sustainability by incorporating the principles of sustainable development with urban design guidelines and process. Different research studies have elaborated on the key principles that should be incorporated into urban design to promote sustainability. As highlighted the major tenets of sustainable development that should be integrated with urban design. These principles include intergenerational equity, public trust doctrine (maintaining environmental diversity and carrying capacity), precautionary principle, intra-generational equity, participation and polluter pays principle. Atkinson and Ting (2002) proposed a framework of transformative sustainable urban design with the following principles: acknowledgement of fundamental ecological patterns and limits, environmental and social restoration and regeneration, seeking better quality of life through live ability, employing integrative and holistic strategies and solutions and recognizing sustainable urban design as a process and product. It is opined that sustainable urban design should be able to provide adequate answers to questions on the aesthetics of the urban form, functionality of the built environment and the sustainability of the social and economic processes.

In a bid to present the sustainable urban design principles in an applicable manner, as elaborated on the key principles and highlighted ten basic tenets of sustainable urban design that are found in expounded literature. These include:

- Stewardship – integrated planning, enhancement through change and town centre rejuvenation;

- Resource efficiency – economy of means, minimal environmental harm, reducing travel/energy reduction and recycling;

- Diversity and choice – variety, permeability, mixed development and hierarchy of services and facilities;

- Human needs – legibility, aesthetics, security, low crime, social mix and imageability;

- Resilience – flexibility and ability to adapt to change;

- Pollution reduction – low pollution and noise, water strategy, climate and air quality;

- Concentration – polycentric city, compact intensification and support services;

- Distinctiveness – heritage, creative relationship, sense of place and regional identity;

- Biotic support – urban greening, open space, biotic support and symbiotic town/country;

- Self-sufficiency – environmental literacy, local autonomy, consultation and participation.

He also noted that the spatial scale of urban design (from local to metropolitan) should be considered in applying urban design principles. In the same vein, scientist suggested three levels of urban design interventions that include individual space, city district and city/conurbation levels. In order to be effective, urban design interventions should have development frameworks generated at these levels. Scientist argued for paying more attention to the development of sustainable neighbourhood since the city cannot be sustainable with unsustainable neighbourhood.

Jabareen (2006) identified seven concepts of sustainable design that are similar to the ones developed by Carmona (2001). His principles are more specific about the issues to be addressed in fostering sustainable urban design. The concepts include:

- Compactness – intensification of built form

- Sustainable transport – design that promotes walking, cycling and transit-oriented development

- Density – high density development

- Mixed land uses – diversity of functional land uses

- Diversity – diversity of land uses, rents and architectural styles

- Passive solar design – reduction of fossil fuel consumption

- Greening – provision of adequate urban green areas

A number of the sustainable urban design principles enunciated above need further research to clarify effect of applying the principles on the urban environment and the direction (increasing or decreasing) of application. For instance, although different authors have tried to document the negative effects of sprawl development, the principle of concentration in form of compact city development is still debatable. As noted that the exact forms and structure that would render the city more sustainable remain elusive and the claims in support of one or the other urban structure are not substantiated. The compactness of the city must be decided with due consideration to the cultural, social and environmental context of the city.

The principles highlighted above are substantive, in line with the criteria, and might not achieve sustainability without sustainable design procedure. The design process and the outcome of design should be sustainable. Scientists suggested frameworks for ensuring sustainability in the urban design process. Scientist identified five aspects of the urban design process that should be integrated. The aspects include substantive, procedural, methodological, policy and institutional and expatiated on the procedural aspect by suggesting a design framework that is within a balanced structure of "Top-Down" and "Down-Top" dialoguing with reflection of various components of the society. The sustainable design process approach consists of five principal stages: sustainable design objectives, sustainable design guidelines, sustainable design statements, preliminary design and sustainable design scheme. This framework has to be integrated with the planning process to promote sustainability of cities. The planning and urban design process should be incremental and adaptive due to the need for flexibility and adaptation to changes in social, economic and environmental contexts. The section below includes further discussion on sustainable design process.

## 9.5 THE INTEGRATED APPROACH TO SUSTAINABLE URBAN DESIGN

As illustrated in the above, there are at least three contending issues/concepts of urban form and design in Saudi Arabia; the heritage of the traditional urban form, modern concept of urban pattern and the emerging and overarching concept of sustainable urban design. Contemporary urban design cannot be based solely on the traditional models as the variables that have contribution to the development of urban development are fast changing and it could be very difficult and impracticable to conceptualize these changes in the light of traditional concepts alone. On the other hand, indiscriminate adoption of modern models has been found to be problematic and incompatible with the traditional city forms. "The total neglect of the traditional forms, and the implications of their meanings and values, will cause us to lose forever our heritage and architectural identity". Sustainable urban design could provide the opportunity of integrating the traditional and contemporary models in a resourceful manner, as some of the relevant elements of both models are embedded in the principles of sustainable design. As noted that the more sustainability principles are applied to design and planning (in Northern Europe) the more these tend to take on traditional forms. However, there should be a framework that lays emphasis on the consideration of traditional concepts for the appropriate integration of these concepts in the principles of sustainable urban design.

The integrated approach to sustainable urban design should have at least three dimensions; substantive, procedural and methodological. The other two dimensions, policies and institutional aspects, are not directly under the control of urban designers. Yet, urban designers need the skills to manoeuvre the institutional framework and also promote the implementation of sustainable policies. The procedural dimension has been highlighted. That is, the procedure should consist of at least five principal stages of objective, guidelines, statements, design and scheme. The integration of sustainability in urban design process is necessary because the process of urban design determines the transformation of the design and the process cannot be divorced from the product. If the design process is sustainable there is likelihood that the design itself will be sustainable and consequently the community will advance towards sustainability. The issue of implementation is very important in the procedural aspect and efforts should be made to monitor and improve the efficiency and effectiveness of the implementation mechanism.

## 9.6 SUBSTANTIVE DIMENSION – A SYNERGY OF PRINCIPLES

The underlying substantive principles of sustainable urban design should be integrated with the traditional values and principles of urban form and expatiated on the underlying principles of traditional Islamic design and highlighted six principles which are fundamental to traditional design. These principles include:

- ◉ Unity – functional and aesthetic forms that expresses an integrated, indivisible whole (unity in space and pattern, in light and colour and in space and form);

- ◉ Openness of space – positive, active spaces interact with negative spaces to express interrelationship between form and space;

- ◉ Simplicity of form and design – use of basic geometric shape and a pure modification and abstraction of geometrical form;

- ◉ Simplicity of structural expression – organic relationship between structural components;

- ◉ Scale – respect for human scale in both the whole environment and particular buildings;

- ◉ Harmony, compatibility and balance – harmonious integration of structure and form; and

- ◉ Privacy – respect individual right to privacy.

The traditional urban and legislative elements which interacted to shape the urban form and spatial structure should also be integrated into the design process.

A set of principles and elements from traditional and contemporary design are selected to be integrated with the principles of sustainable urban design to engender better urban form and design . As shown in figure 6, the sustainability principles include mainly the concepts and some issues that were not explicitly mentioned in his framework.

For example, resilience to climate change and natural hazards has become an important issue in recent time due to global environmental change. Human needs such as social relationship are also very important.

The process issues such as public participation and institutional framework are not substantive but they are also important. So, they are shown in red boxes outside the substantive triangle (Fig. 6). Some of the issues feature in more than one set of principles to show the overlap between some of the contending principles. Some challenges might arise from the issues that are conflicting among the principles. For example, promoting urban greening might conflict with conservation in a desert environment. Maintaining green areas require a lot of water and it might impact on energy demand. Also, it might be difficult to achieve high density neighbourhood with design of courtyards. The principles highlighted in this study are not conclusive and other principles could still be valid and relevant. In essence, the paper proposes that different principles of traditional, modern and the overarching sustainable urban design should be applied in an integrative approach to achieve better and liveable cities.

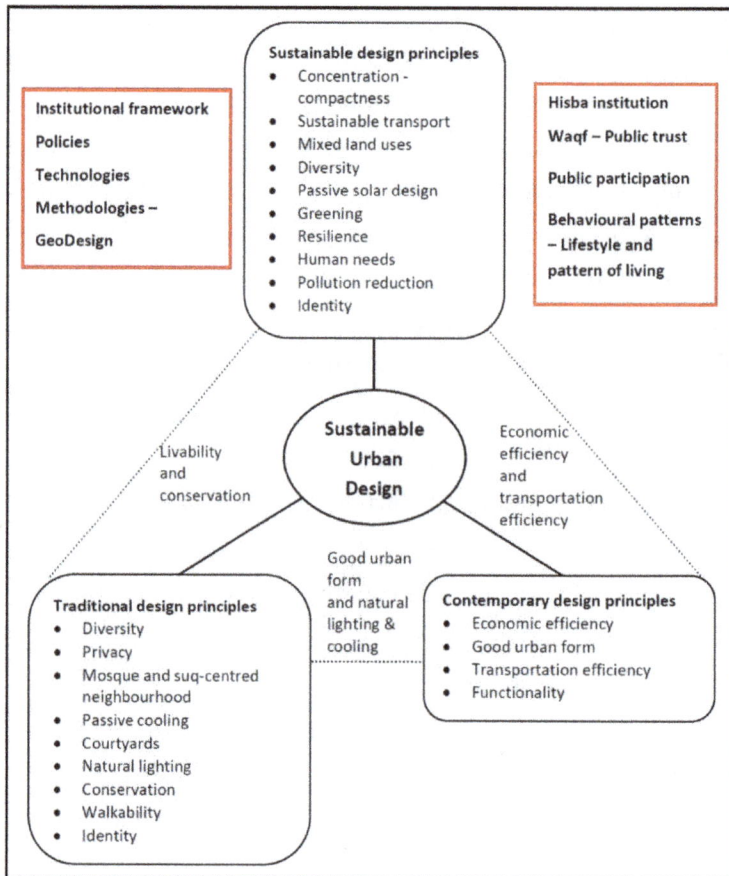

**Fig. 6: Conceptualization of the integrative theory approach**

# 9.7 METHODOLOGICAL DIMENSION–A CASE FOR GEODESIGN

The methodological dimension involves the utilization of different analytical, descriptive and modelling techniques to fully grasp the complexity of different factors involved in sustainable urban design. No single method is capable of analyzing the components of urban systems in a comprehensive manner. Thus, there is the need to integrate different methods of analysis with a view to further understand and model the urban system. Efforts have been made in this direction with the development of space syntax, cellular automata, GIS and the collaborative planning support systems. It has been noted by Batty et al. (1998) that the advances in computer models and information systems have hardly been fully utilized in urban design. It is highly pertinent now to find ways of utilizing the powerful potentials of different computer models and information systems to support urban design. Scientist identified about four ways in which urban system can be represented (by the information systems) at the level of urban design. These include the representations of socio-economic, functional, behavioural and physical information. Virtually all the information can be stored and analyzed digitally by the current level of technology. There is just the need to fully integrate the available information systems and make them amenable to supporting sustainable urban design process and product.

Efforts towards the integration of information systems for urban design have led to the emerging concept of GeoDesign. It is the adaptation of geography, geographic information system and other information systems in a synergetic way to support urban design. That is, "integrating geospatial technologies into the design process with the goal of living more harmoniously with nature" (Artz, 2010). Goodchild (2010) defined GeoDesign as "planning informed by scientific knowledge of how the world works, expressed in GIS-based simulations". Apart from utilizing the sketching and simulation capability of GIS (Goodchild, 2010), with geographic reference, GeoDesign uses web and visualization technologies to enhance collaboration and stakeholders participation during the design process. The broad idea is to have all design related technologies, such as computer-aided design (CAD) and building information modelling (BIM), integrated into GIS to be able to design in a spatially aware environment.

GeoDesign has been demonstrated to include the following essential elements

- Sketching – drawing proposed designs or plans
- Spatially aware simulations – modelling different systems (environmental, economic and so on) and how they will respond to proposed design in terms of impacts and change (with geographic reference)
- Fast feedback – supporting collaboration
- Iteration – trying and visualizing different alternatives

◉ 3D visualization – presenting design alternatives and impacts in three dimension

The capabilities of GeoDesign make it a valuable tool for urban designers in promoting sustainable built environment. Urban designers need capable analytical, modelling and visualization tools to synthesize the varying issues of urban complexity, climate change and human social needs.A model by Dangermond (2009) successfully highlighted the importance of GeoDesign in design process (Fig. 7). The model incorporates the design process with the elements of GeoDesign such as designing, sketching and geo-accounting. It showed that Geodesign could be a veritable box of tools that enables "a rapid and adaptive process for creating a sustainable future" . In terms of usage, most of the applications of GeoDesign are still in Europe and North America but there is a growing adoption in other countries. In Saudi Arabia, GeoDesign is gaining momentum because GIS technology (backbone of GeoDesign) is already being used in the kingdom to solve different geospatial and design problems.

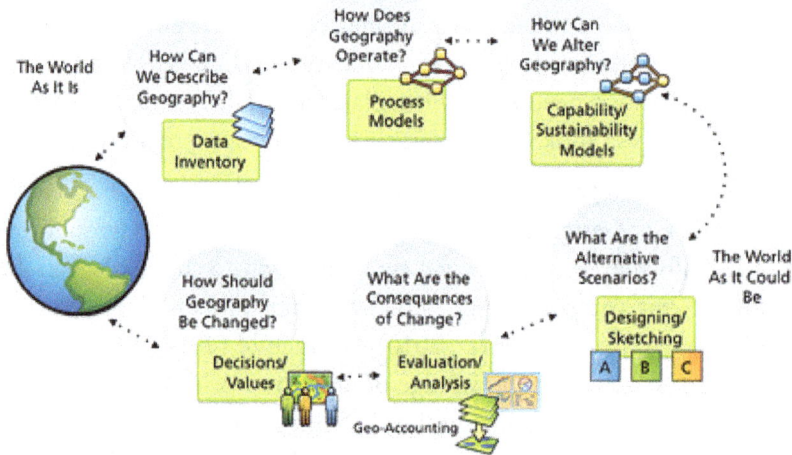

**Fig. 7: GeoDesign in the sustainable design process (Source: Dangermond, 2009)**

**Fig. 8: Building facade with window screen similar to traditional mashrabiyah (Source: A. T. Service – Wikimedia commons)**

These highlighted constitute a very bold start in embracing the principles of sustainable urban design in the Kingdom. The challenge is in replicating such projects across the Kingdom and making the adoption of sustainability principle the norm. It is noted that there few promising traditional designs here and there but there is no coordinated effort to implement them on a general level. Another challenge is the ability to change the behavioural pattern of the populace.

# 10

# ROLE OF URBAN DESIGN IN DEVELOPING COMMUNITIES

## 10.1 INTRODUCTION

Historically, cities and society developed and flourished in an almost symbiotic manner. The Latin word for city is civets, from which the words civilization and citizenship are derived. Despite being home to a minority of the population prior to the industrial revolution, cities dominated their surroundings and exerted immense influence in all spheres of human endeavor. Although there was a tense division of wealth and power between the countryside landowners and the city-based merchants, over time the latter prevailed. As well as being seats of power, wealth and knowledge, cities have also been catalysts for social change and revolution. They have been the source of most of the lasting developments that underpin human freedoms. Arguably cities are the birthplace of democracy.

*The primary focus of the city is its people, operating at a human scale, rich in symbolism of spaces and places for social interaction and the daily business of life. The relationship and hierarchy of spaces are vital to an urban environment if it is to bring meaning to people's lives. The relationship and hierarchy of buildings is no less vital. Two building traditions in cities have been identified, one Focal and the other Contextual. Focal buildings should represent society and need to re-establish their symbolism. At the same time, a new shared architectural language based on specific localities must be devised for contextual buildings to return to a subordinate role. Places can and should enhance their communities by providing landmarks in time and space that are perceived as touchstones for the past and present.*

The essence of cities is that they contain a myriad of diverse and intense connections and activities, where people live, work, shop and play, meeting needs of economic production and social reproduction. They bring together people from many different backgrounds and cultures. This can be purely in terms of physical proximity but can also create the space for a ferment of ideas, styles and activities. They are centers for many cultural forms. Without romanticizing their history, which has its grim share of oppression, hunger, crime and pollution - cities have been the driving force for innovation, social improvement, cultural activity and diversity.

As well as density of population and networks of activities, cities have public buildings and spaces for government, organized religion, education, commerce, social interaction, cultural events and public services. These buildings and spaces play an important role in providing a focus for communities. They offer emotional attraction for both citizens and visitors, embodying political and cultural significance, and landmarks in time and space. They link the past, present and future, become reassuringly familiar to local people and stimulating for visitors. Lozano (1990) argues that the city is a realm with a high level of culture linked with the most civilized expression of social behavior. Mumford (1966) describes the city as the most advanced work of art of human civilization.

Traditional cities have had complex spatial layouts reflecting the multiplicity of human exchanges. They have been alive with the richness of patterns and symbols that fulfil many psychological and spiritual needs. For example, the sense of enclosure and spatial definition provided by medieval walls satisfied more than just a need for defensive protection. They also provided psychological stimulation and physical comfort. The need to pattern human surroundings is as valid today as it was in medieval times. Their ideas about legibility are based on a vivid and integrated physical setting that can provide the raw material for a symbolic and collective community memory. The layout, landmarks and public spaces all contribute to each city's distinctive sense of identity.

Urban space has always been the place for the community rather than the individual, where activities are representative of the distinctiveness of that settlement. It was where the framework of society was debated and formulated, and where economic activity took place. It believes that modern cities have lost sight of the traditionally understood importance of urban space. Of all the types of urban space, the square is most associated with the values of the society that created it – the agora, forum, cloister, Muslim courtyard, are all examples. Public spaces are not necessarily benign; many have been used for martial rituals or public execution, yet they also have a democratic tradition that allows access for the community.

## 10.2 FROM CENTRES TO SENSELESS

The last two centuries have seen a dramatic transformation in cities from being largely of walking distances and embedded within a neighboring countryside to widespread urban sprawl. As a consequence, the urbanareas of today while still having a dense concentration of population, have lost much of the essence and character of the city.

The combined actions of economic power and planning have undermined the importance of distinct spaces and landmarks that established the character and spaces of cities. Many urban patterns and traditional connections have been weakened or lost, slashed by mega redesigns that ignored centuries of evolution. Cities have been scarred by major road networks, which occupy large areas of land, fragment and blight neighborhoods destroying local social interchange and disconnect travelers from their surroundings. Sprawl, traffic, zoning and major redevelopments have destroyed the fabric of buildings and spaces, often replacing diversity with large single-use structures that have a hostile and imposing presence.

**Fig. 1: Model for mid-20th Century City Centre Development**

In the process of modernization, urban communities have lost the richness of patterns and symbols that made each city distinct. One of the experiences of the 20th century has been that planning tended to drill all places into a similar type while development based on market forces has created a meaningless jumble of buildings, lacking all sensory stimulation, sense of place or even idea of location. The concept of citiesas being the setting of culture and civilized behavior became increasingly fragile.

In modern cities, central places are rare, leaving only a spatial vacuum between buildings. This approach tends to be symptomatic of thebureaucratic non-city where urban spaces are created by default if theyhave not already been replaced by roads. In contrast, good urban design recognizes that there is an advantage in emphasizing some elements andsubordinating others. The effect of the symbols of society is lost if they arescattered aimlessly around the city. An impression of unimportance is suggested if they are found in ordinary streets or next to the by-pass. Many of the buildings that now dominate cities such as office or residential blocks, retail and leisure complexes would formerly have been contextual developments. In the 20<sup>th</sup> century modern city, they appear as landmark buildings such that the order of built form has been turned upside down. Many public buildings have been rendered insignificant, losing much traditional symbolism in the process. Landmarks are no longer socially, politically or even economically significant and rarely visually, culturally and spiritually uplifting.

Today's modern landmarks reflect the values of commercialism, where offices and retail units have replaced the library and the town hall. This reversal damages the fabric of the city as such buildings do not connect with the social or physical fabric, nor reflect the identity of a particular city and its community. As argued that when there is a physical loss of cuteness in favors of huge bland conurbations, there is a parallel loss of citizenship. Active citizenship is discouraged and replaced by the role of consumer or, at best, passive voter. Power is concentrated in large companies and government tends to act in their favor. Recent trends in global economics and telecommunications have produced the claim that place no longer matters.

## 10.3 RENAISSANCE

The last few years have seen a re-emphasis on the importance of cities with encouragement of city living, calls for an improvement in the quality of urban design, and support for public space. In this way, urban design can be connected to the social and environmental networks in which buildings and spaces exist. Designing buildings in scale with the traditional pattern of a city could be a first step towards regaining an urban community. This is not to denigrate the desirability of a proper system of planning. Indeed, the advantages of a sympathetic plan, prepared with forethought and care to provide for the needs of the community are self-evident. Yet, despite a long history of planning on behalf of the community, one of the problems of the early 21st Century is that there has been a great decline in the public realm. We may be richer as individuals but as citizens we are getting poorer. There has been a retreat into the private realm - with emphasis on personal comfort, consumption and security. A response could be that development in different areas of a city needs to be guided by a planned framework. Development cannot be based on unconscious and accidental character but must reflect the rules of conscious and ordered design. However, each framework needs to be specific to a locality. Topography and other natural features that are unique to a particular place should be integrated and emphasized.

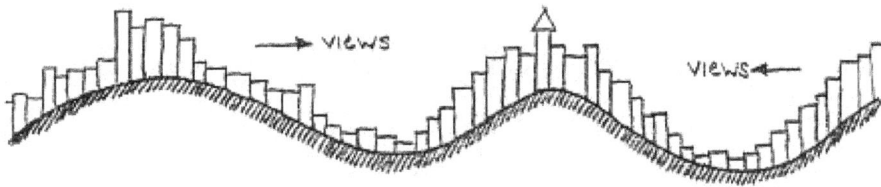

**Figure 2 Topography and Building Heights**

For instance, keeping buildings at valley bottoms to the same orsmaller scale to those at the tops of hills actually emphasises the topography. Other contrasts in response to topography are clearly evident in the following illustrations:

TRADITIONAL BUILDINGS                    MODERN PLANNING

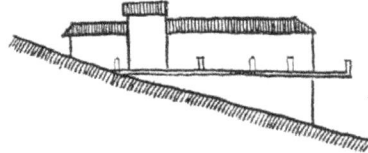

**Fig. 3: Built Responses to Topography**

Buildings related to topography – small units capable of stepping down a hill creates a varied and interesting streetscape Large boxes, incapable of dealing with the topography except by introducing artificial levels and ramps resulting in a monotonous streetscape

If traditional methods were taken into account in modern planning, the result would be a city plan thatencourages parks, gardens squares and streets, all defined by groups ofbuildings. In this analysis, there are two distinct building typologies. The first is associated with professional design and the second with thegeneration of human habitats (Giddings 1999). Buildings of professionaldesign should be recognised by styles of high culture and should be reserved for buildings that have symbolic functions in a particularcommunity. For example:

| Government | castles, palaces, parliaments, city halls |
|---|---|
| Religion | temples, cathedrals, churches, mosques |
| Public Facilities | museums, theatres, exhibition halls |
| Law | courtrooms |
| Health | hospitals |
| Education | universities |

A few exclusive residences may also be included but these are usually part of the political, social and economic power structure of the community.

Design solutions in this genre should originate in the rules of style, be impressive, prominent and aimed at achieving masterpieces. In this context, innovation is an essential ingredient. Moreover, as the buildings are representative of the wealth and power base of a civilisation and atleast in theory, designed as a whole, the wealth associated with them should reduce delays and make them available to the community as they are needed.

The second type is more to do with context, unity, harmony, etc. Often the buildings appear in what is known as the vernacular. They are essentially private and offer more commonplace uses, such as: residence, employment, leisure and shopping. This appears to be the basis for a sound framework but there are difficulties with it. Invariably, so much has been overlaid upon a city that any local tradition occupies at best a minorityof space and is therefore of dubious relevance. Secondly, an international style of building may have subsumed any traditional methods. This raises questions as to whether other styles might also be legitimately included.

Thirdly, there may be a problem where new uses demand a scale and/or type of space not generated by a local building tradition. Shopping is one use that has tended to outgrow its traditional accommodation – with the advent of supermarkets, superstores, department stores, retail warehouses, etc. Yet, regardless of these difficulties, a policy that expresses the difference between focal and contextual buildings will greatlyassist in restoring fascinating and functional urban environments. In many successful situations, each focal building is related to a place of assembly, outside its main entrance. This introduces an important concept that the entrance to each focal building should lead onto a public square. Also, the position of the focal building tends to determine the direction of the square. Some pioneers go further to suggest that in deep plan squares, the focal building should have vertical emphasis, eg a church, whereas in wide plan squares, it should have horizontal emphasis, *e.g.* city hall.

**Fig. 4: Deep and Wide Plan Squares**

Contextual buildings perform two roles. First, they provide the frame for the focal buildings. Secondly, they define and contain urban space. It is the contextual buildings that should form the squares and define the streets.

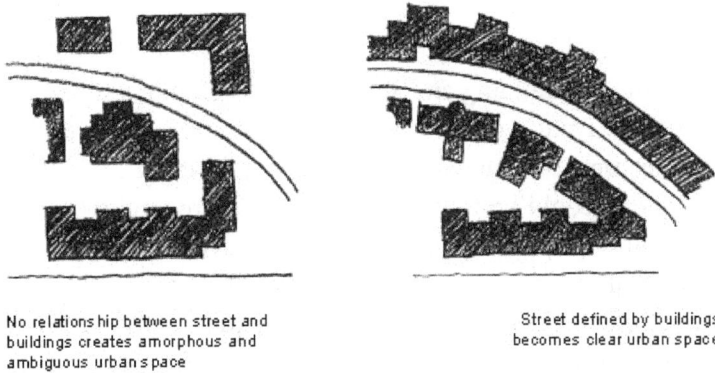

No relationship between street and
buildings creates amorphous and
ambiguous urban space

Street defined by buildings
becomes clear urban space

**Fig. 5: Relationships between Street and Buildings**

Thus, there is a need for visual continuity, even if the buildings are not physically attached in a continuous manner. The nature of these rows of contextual buildings can be expressed as Architectural Frameworks, two examples of which are illustrated below:

| Assertive Architectural Framework | Passive Architectural Framework |
|---|---|
| • Informal | • Formal |
| • Greater variety of styles | • Limited number of styles |
| • Greater range of materials | • Limited range of materials |
| • More elaborate facades | • Simple elevations |
| • Emphatic changes in building line | • Minimal changes in buildingline |
| • Raised skyline | • Little skyline interest |
| • Narrower frontages | • Broader frontages |

**Assertive Architectural Framework** *example: Amsterdam, Holland*

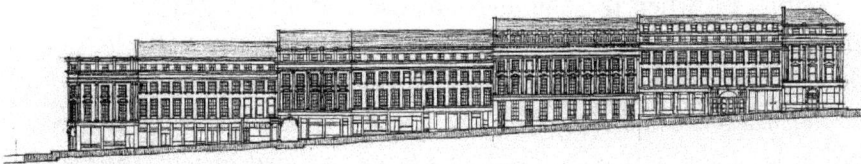

**Passive Architectural Framework** *example: Newcastle, England*

**Fig. 6: Architectural Frameworks**

# 10.4 INTERACTION WITH CITY CENTRE SPACES

It has already been established that the entrance to every focal building should lead onto a public square. If each of these buildings also displays a distinct attribute of the society it represents, then each square marks the arrival at that symbol of society. Individually, every pairing of symbolic building and square can have quite a dramatic effect on the psyche of the citizens. Traditionally, clusters of urban spaces have been such frequent phenomena that they were considered the rule and single public squaresas the exception. An objective therefore, could be to create groups of interconnected places rather than isolated statements. It is fun to meander from space to space and place to place but people need different kinds of movement and a multiplicity of routes from the very direct to a variety of options. The special effect that results from walking about from one square to another in a cleverly grouped sequence is that our reference points change constantly, creating ever-new impressions.

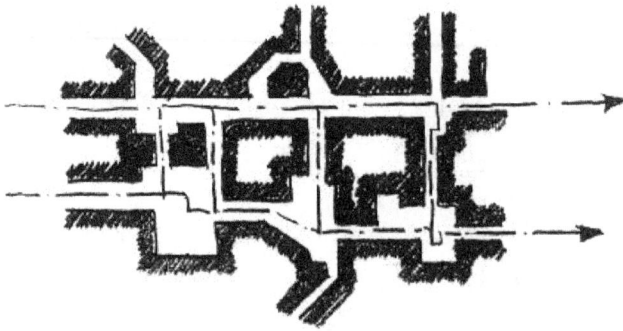

**Fig. 7: Different Kinds of Movement through Spaces**

Human life, activity and culture depend on the environment. The layout of a city can encourage social interaction or act as a form of social control. In the latter case, if spaces for assemblies, rallies, events, entertainment etc., either do not exist or are so ill-conceived that they are not welcoming, then the inhabitants will live their lives independently or at best in small groups. Certain political organisations delight in this form of social control but it does not suit mature European societies. In Stuttgart for example, social interaction in public spaces is greatly encouraged. It is suggested here that the kind of ethos, which enables useful public space to thrive, strengthens the identity of place and the deep psychological connections of the community with the place. This approach becomes all the more significant when the buildings are also considered. This city centre contains some powerful symbols of society. The Rathaus (Town Hall), symbol of local government and political order, stands with its main entrance opening onto Markt-Platz – the physical and metaphorical focus of the city. The Altes- Schloss (Old Castle) is a symbol of history, longevity and past conflicts resulting from a less developed society. The Justiz Ministry symbolises law and order, and the Stiftskirch represents spiritual fulfilment. These three buildings define Schiller-Platz, only second in importance to Markt-Platz in the spatial hierarchy. The buildings around this square are the only ones in the centre to pre-date the 20th Century a reminder that they do not represent temporary or transient values. The square itself pays homage to the great German poet, indicating the significance of the Arts. Along Dorotheenstrasse is the Markthalle, symbol of trade and sustenance. The proximity of the church and market illustrates the two sides of human need ie spiritual and physical. The proposition is that the community is psychologically re-assured by the clarity of these symbols as they move through the public spaces. There is a permanence of civilisation that is associated with the layout.

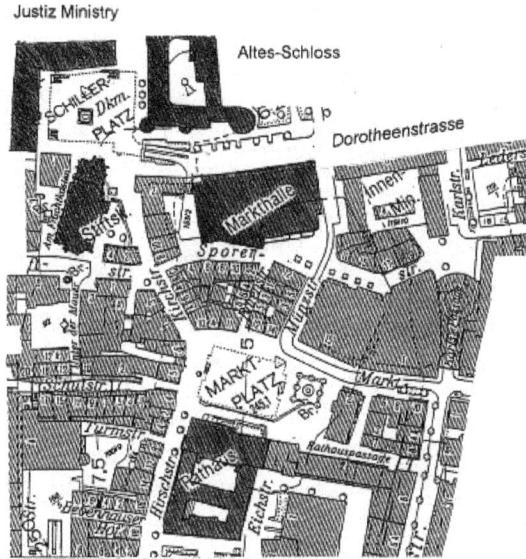

**Fig. 8: Stuttgart City Centre**

Often the best way to strengthen the center of cities is to support the existing local people, business and activities local culture and festivals. These enhance the quality of the environment, encourage walking, support public places and buildings, and design for people. While this may not appeal to the property developer or the politician keen on the big change; it may take time but is organic, connected to the city and more likely to last. This all contributes to distinctive places, rich in diversity and activity.

The essence of cities is close connections and dense activity. Pedestrians connect and interact with their surroundings. Walking and travel on public transport provide the opportunity for chance meetings, the serendipity that enriches life, while people sealed in a car are isolated from society. City centers can be designed for walking among a diversity of activities including houses, workplaces, shops, pubs and cafes, places of worship and public spaces. As Healey (1998) has pointed out, place does matter contrary to the claims of some neo-liberals and advocates of globalization. Sense of place generates the soul of the city and the principles of a renaissance in city design celebrates that experience. Cities need their unique sense of being a distinct place. Active and engaged citizenship in planning and developing the urban form to meet local needs will begin to shift the trajectory of cities away from the maximization of profit and consumption towards the well-being of the community. Society needs to challenge the view that the market economy is the only form of socio- economic organization that can provide for a society. Cities can be places where the interaction of citizens enables them both to meet their own needs and aspirations and those of the widercommunity for this and future generations.

The primary focus of the city is its people, operating at a human scale, rich in symbolism of spaces and places for social interaction and the daily business of life. The relationship and hierarchy of spaces are vital to an urban environment if it is to bring meaning to people's lives. The relationship and hierarchy of buildings is no less vital. Two building traditions in cities have been identified, one Focal and the other Contextual. Focal buildings should represent society and need to re-establish their symbolism. At the same time, a new shared architectural language based on specific localities must be devised for contextual buildings to return to a subordinate role. Places can and should enhance their communities by providing landmarks in time and space that are perceived as touchstones for the past and present. There is little doubt that a community uplifted by its environment is more socially adjusted, more economically prosperous and more optimistic about the future.

# 11

# CONTEMPORARY URBAN DESIGN: THEORY AND PRACTICE

## 11.1 INTRODUCTION

First things first: let's not assume that we all share the same definition(s) of urban design. After all, the discipline has undergone many changes since it was founded in our modern era as part of architecture's embrace of the city as a site for design practice and theoretical discourse. In this first unit, we will briefly review the nineteenth century origins of urban design in the modern age, and explain its main concepts as they changed through the middle decades of the twentieth century. We will then examine how this panoply of ideas about cities underwent a series of transformations and recapitulations to become the ideas that define urban design practice today, in the second decade of the 21st century.

The term "urban design" as used in this course draws on these historical precedents, but also includes more recent understandings of environmental issues and the social dynamics of places. In straightforward terms, urban design means "the art of making places for people". More specifically, urban design ...draws together the many strands of place-making – environmental responsibility, social equity and economic viability, for example – into the creation of places of beauty and distinct identity. Urban design is derived from but transcends related matters such as planning and transportation policy, architectural design, development economics, landscape [design] and engineering. [It] is about creating a vision for an area and then deploying the skills and resources to realize that vision.Although these are English definitions, they apply equally to American practice, continuing the intertwining of Anglo-American urban design theory and practice that has typified the discipline for the past 150 years.

## 11.2 THE EVOLUTION OF URBAN DESIGN

While the design of cities has been practiced for millennia in different cultures, the term "urban design" as we understand it today in the 21st century was first coined in America during the 1950s. In 1956, José Luis Sert, Dean of Harvard's Graduate School of Design and a pupil of modernist master architect and urbanist Le Corbusier, convened the first Urban Design Conference at Harvard and set up the first American urban design program at that university (Shane: 63). One year later, in 1957, the American Institute of Architects set up a committee on urban design (Rowley: 306). Other versions of the profession's origins note the University of Pennsylvania's Civic Design Program begun in 1957, and place the date of Harvard's urban design program at 1960.

All three accounts of this subject's evolution oddly omit mention of the founding of the very first Department of Civic Design at the University of Liverpool in England half a century earlier in 1909. This historical oversight is surprising, especially as the theory and practice of city design in the U.K and the U.S. followed very similar and overlapping trajectories in the late nineteenth and early twentieth centuries. Indeed, American urbanists consciously followed, and quite often improved on English precedent. This period of urban development up through the late 1920s has left a legacy of sophisticated "streetcar suburbs" in cities and towns across the United States. These so-called "traditional" neighbourhoods derive their name from their use of historical types of clearly defined public infrastructure -- e.g., connected streets, short blocks, civic squares, plazas, parks, and alleys – and by the way these spaces are woven together to create the fabric of everyday residential and commercial life.

The ideas upon which current U.S. urban design is based today have direct connections back to these concepts of traditional urbanism that were taught and practiced a century ago. However, what makes the historical narrative confusing is that these ideas – ones that shaped American cities in the decades of the late 19th and early 20th centuries -- were soundly rejected, even ridiculed, by urban theorists and practitioners during the Modernist period. Between the 1920s and the 1980s, the decades we now broadly characterize as "Modernism," we began to think about cities and how to design them in a whole new way.

But now, by the second decade of the 21st century, urban designers have intentionally discarded those once-dominant modernist concepts and returned instead to a version of America's traditional urbanism, updated to meet new concerns about sustainability and resilience. If this sounds confusing, let 's try to unravel this twisted skein of ideas.

## 11.3 THE SHIFTING LANDSCAPE OF URBAN DESIGN

Urban design techniques in the years around the turn of the nineteenth and twentieth centuries were heavily influenced by the fledgling discipline's twin lineage from architecture and landscape architecture, and the predominant method was the physical design "blueprint," or master plan. In most cases these master plans included some form of neighborhood center, often based around a commuter rail or streetcar station, and including housing and commercial uses in buildings framing public space. This focal area was then surrounded by a connected network of streets lined with a variety of different types of housing. Parks and other green spaces were integrated into these neighborhoods, and attention was paid to the adequate provision of sunlight into dwellings, to avoid the dour and grimy spaces typical of the industrial cities of the period. These new suburban developments became known as "romantic garden suburbs" -- largely due to their origin in 19th century England as middle-class versions of the country estates of the aristocracy – or "streetcar suburbs" because of their use of streetcars as the main means of public transportation. The period from the 1890s to the 1930s proved to be the heyday for this type of urban development.

At the beginning of the twentieth century in America the City Beautiful movement provided a grand, neoclassical vision of civic design that further informed the status of civic space and civic architecture at the heart of communities. Meanwhile in Britain, Raymond Unwin's massive tome Town Planning in Practice (1909) became a seminal text that wove all these strands together into a rich historical, theoretical and practical manual for designers of that period. This book is back in print and still contains a wealth of useful knowledge for today's urban designer.

A good example of this kind of development can be found at Forest Hills Gardens in Queens, New York, designed as a "model town" in 1909 by architect Grosvenor Atterbury and landscape architect Frederick Law Olmsted Jr. for the Russell Sage Foundation.

This development illustrates perfectly both the design principles noted above, and the transatlantic influences; both Atterbury and Olmsted looked for inspiration to the contemporary English developments of Letchworth Garden City, (1903) and Hampstead Garden Suburb, (1907) in north London, both designed by Barry Parker and his brother-in-law, Raymond Unwin (see Fig. 1). The design quality achieved by Atterbury and Olmsted led to this suburban development being dubbed "the first Garden City in America".

This pattern of urban design and planning remained mainstream American practice until the end of the 1920s. At that time much suburban development slowed due to the onset of the Great Depression, and development patterns began to change with the increase of individual car ownership and the consequent decline of public transit. No longer was it important to construct tightly organized, mixed-use and walkable communities. The private automobile allowed the elements of city life to be more widely spaced apart, but still quickly accessible by car.

**Fig. 1: Hampstead Garden Suburb, London, begun 1907. Master plan by Barry Parker and Sir Raymond Unwin. The Free Church (at end of axis) by Sir Edwin Lutyens (1908-10). Note how the urban designers place a significant civic building to close the axial view and intensify the "sense of place."**

Two key neighbourhood plans from the late 1920s illustrate this shift of design thinking: Clarence Perry's Neighbourhood Unit plan from 1929 (as part of the New York Regional Plan), and Radburn, New Jersey (1928- 33), by architect/planners Clarence Stein and Henry Wright, together with landscape architect Marjorie Cautley. Neither plan involved public transit – the private automobile was in the ascendency at that time, while public transit was declining -- and both designs placed emphasis on safeguarding the

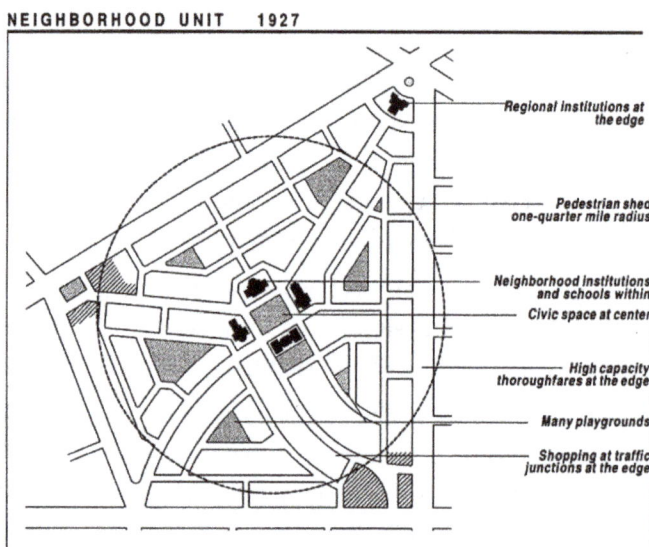

**Fig. 2: Clarence Perry's Neighborhood Unit. The circle denotes a 5-minute walk, approximately ¼ mile.**

residential areas from fast moving traffic. Parry's plan was the more "traditional" of the two, with a connected network of small-scale streets around a local community hub, and his ideal neighbourhood was still conceived as an entity embraced within a five-minute walk (1/4-mile radius).

However, faster, through traffic was channelled along the edges of the neighbourhood, and a larger urban area could be created by combining several neighbourhoods together on either side of the main arterial roads (see Fig. 2). The traditional, still walkable layout within each neighbourhood was likely influenced by Parry's time spent living in Forest Hills Gardens.

By contrast, Radburn placed primary emphasis on the almost total separation of pedestrians and vehicles, with cars kept out of the main pedestrian, landscaped areas. These landscaped "linear parks" were crisscrossed with pedestrian pathways, which were intended to link together through underpasses beneath the major highways, leading pedestrians and cyclists safely to commercial and cultural clusters of uses (see Fig. 1.3).

All the local streets ended in culs-de sac, but this created an unfortunate dichotomy between the " front" doors of the homes facing the public parkland and public footpaths, and the "back" doors opening off private back yards and car parking areas. (The design was never completed due to the bankruptcy of the development organization in 1933 during the Great Depresion).

Increasingly most visitors arrived by car, drove down the cul-de-sac, parked and entered through the private rear garden and the "back" door. This arrangement inevitably led to confusion regarding which entry was the public "front" for visitors, and the traditional distinction between public fronts and private backs was scrambled. This led to the "front" door facing the park and its pathways falling into disuse; in some instances, it was even blocked off with furniture for extra wall space within the home. This was evident to the author during a visit in the 1970s to an English housing development in Runcorn New Town that was based on what a U.K. government report later called the "failed" Radburn layout (Communities and Local Government Committee, 2008).

**Fig. 3: Radburn, NJ. 1928. Clarence Stein, Henry Wright and Marjorie Cautley.**

Radburn was a well-intentioned experiment in new patterns of residential development suited to the automobile age. Its focus on pedestrian safety from cars was timely, and the incorporation of landscaped parkland as an integral element of the design was noteworthy, but the layout of separated roads and pedestrian paths led to this major confusion of fronts and backs, and public and private spaces.

In this context, Radburn was part of a major shift in design thinking about city form and function. Instead of using "traditional" types of urbanism rooted in the western tradition of city building – where buildings came together to frame shared networks of "joined up" public spaces such as streets, squares, plazas, courtyards parks and alleys, the new, "modernist" concepts prioritized the idea of buildings set apart in lush landscape for pedestrians, with vehicles segregated into a separate, functional system of access roads. In essence, the concept of cities formed by shared public spaces enclosed and defined by building façades was rejected and replaced by the city of separated objects existing in a "flow" of universal space. At a stroke, thinking of cities as a series of connected and defined spaces was replaced by conceptualizing cities as a collection of objects sitting in space and serviced by a separate system of roadways.

## 11.4 THE RISE AND FALL OF MODERNIST URBAN DESIGN

This tectonic shift was not one of style and aesthetics alone. Indeed, its origins resided deep inside a humanitarian desire for social and physical reform of the harsh conditions typical of the 19th century industrial city. Under the intellectual leadership of a new European avant-garde in the 1920s, featuring designers such as Le Corbusier, Walter Gropius, Ludwig Mies van der Rohe, Ludwig Hilbersheimer, and the artists and architects at the Bauhaus, architects passionately sought to rid society of the environmental and social evils of the cramped, polluted and disease-ridden industrial city. This was the setting where teeming crowds of workers lived miserable lives, crowded into dismal and unsanitary slums. In place of this old, corrupted Victorian city, modern architects envisioned a bright, new healthy environment, full of sun, fresh air, open space, and greenery, where bold new buildings, free of the trappings of archaic historical styles, were sited apart in a bounteous, sun filled landscape. It was a terrific utopian vision and a fulfilling professional mission.

The key summary of these new city design concepts was provided by the Charter of Athens, published in 1942 under the auspices of Le Corbusier, and which codified the modernist view of urbanity. This emerging new doctrine about cities had been formulated originally in 1933 by CIAM, a movement that was founded in 1928 as means of propagating the agenda of modern architecture. Specifically, it sought to unite a series of disparate architectural experiments into an international movement with common intentions. As part of this unification, architects sought cohesion around the building style that had emerged strongly at the Weissenhof exhibition in Stuttgart, Germany, and which we know today as the "International Style" (see Fig..4).

The original 1933 version of the Charter was formulated during the 4th Congress of CIAM. As a relief from the political tensions in Europe in the 1930s due to the rise of fascism in Germany and Italy, the conference itself was held aboard the steamer S.S.

Patris II as it sailed across the Mediterranean from Athens to Marseilles. The crusading document we know today is in fact a substantial and subsequent rewriting of CIAM IV's original maritime proceedings. The mildmannered technical language of the original notes, Les Annales Techniques, was transformed by a series of working groups, influenced heavily by Le Corbusier, into a hard-hitting, dogmatic manifesto that eventually appeared in 1942 under Le Corbusier's sole authorship (Gold; 1997).

**Fig. 4: Duplex housing at the Weissenh of Siedlung exhibition, Stuttgart, 1927. A clear exposition of the "International Style."**

The Charter narrowly defined the modern city under four categories – the "Four Functions" of Dwelling, Work, Recreation, and Transportation -- each with its distinct and separate location and urban form. A fifth heading briefly discussed historic buildings and suggested it was appropriate to conserve buildings if they were true remnants of the past. However, the tone of the document implied that no avant-garde architect or planner associated with the modern movement could or should allow these past cultures to interfere with the grand work of making the new city. The premise of the four functional groupings was that each category of building would be optimized within its own parameters, without any compromises from other uses. Absent from the text of the Charter was any meaningful discussion of the social, economic, or architectural character of existing residential or mixed-use neighbourhoods; those "softer" areas of concern did not fit well with the hard-edged functionality of the new theories for making the "brave new world."

The Charter's rhetoric was powerful, however, and its vision was compelling in its distilled abstraction of human functions. The urban ideas enshrined in the text became guiding principles and doctrine for many architects and planners involved in rebuilding British and European cities after World War Two. Moreover, these same ideas were transplanted into American practice in the late 1940s and 1950s by the many European architects, planners and intellectuals who fled fascism and persecution, starting new chapters of their professional lives in the USA.

Within the new vision of urban form that grew from the Athens Charter's Four Function model, the traditional street was singled out for special disdain. Indeed, streets in old slum areas of cities were truly awful – fetid and filthy warrens of squalor. But soon this revulsion extended to all streets, even the charming streets of medieval cities and the grand boulevards of Paris. Le Corbusier famously derided medieval streets as primitive "donkey paths," and now buildings were to arise like "Towers in a Park," sparkling in ubiquitous sunlight. Street networks were now dissolved and bifurcated into access roads and high-speed highways. These theories soon translated in practice. Cities all over America pursued programs of massive land clearance and rebuilding, separating their old, "mixed-up" cities into "clean" and separated districts, each categorized by a different function: housing here, offices over there, and shopping in a third location.

To connect all these separated areas together, major new road building projects carved through cities, demolishing everything in their path. Architects, planners and engineers alike were energized by the quest to build cities anew, and in so doing swept away all the unwanted detritus of past eras.

However, while the theories developed in the heady days of the 1920s painted a grand and compelling utopian vision, the implementation of these concepts in Europe and America during the decades after WW II varied enormously; a tangible gap was revealed between the promise of the utopian vision and "reallife" achievements on the ground. In America, it was often poor African-American neighborhoods that were targeted for clearance, with few, if any plans for rehousing, and the much-vaunted "Towers in the Park" all too easily became "towers in the parking lot." By the 1970s, the planning and design philosophies of the modernist agenda were being severely questioned by the public. Planners and architects first took a defensive position. They suggested the bleak urban environments people were complaining about were simply the result of the great visions of the masters being interpreted by less talented pupils, but increasing popular discontent, particularly against racially biased programs of urban renewal in America, gradually made the modernist position untenable.

The uniformity and abstraction of modernist, "International Style" buildings puzzled and dismayed a public used to a richer and more conventional architectural language of historical detail and imagery, even in the most modest of structures. Over time, redeveloped urban areas bred a form of distaste and antagonism among residents who lived and worked there. In particular, the large tracts of semi-public space between the isolated buildings that were the norm in most urban redevelopments, from the 1950s through the early 1970s, gave rise to unforeseen and uncomfortable ambiguities about social behaviour.

This open space between buildings was prescribed by modernist doctrine to allow universal access to sunlight and greenery, but in practice this space was neither truly public nor private, and its consequent lack of spatial definition blurred boundaries and territories, raising issues of control and management, and ultimately of crime and personal security (see Fig. 1.5). Few people living in the large, modern housing redevelopments of slabs and towers favoured by modernist theory felt safe or comfortable, or felt sufficient ownership of the open spaces around the new buildings to help take care of them. The list of failings in urban renewal and redevelopment schemes grew to such length and seriousness that ultimately it was impossible to treat these problems as teething troubles or poor applications of visionary ideas by less talented designers. As urban historian John Gold has pointed out, a movement predicated on functionality as a core belief could not withstand criticism about its dysfunctional consequences.

**Fig. 5: "Tower block" public housing in Newcastle, UK. 1967. Undefined open space between buildings became vandalized and a breeding ground for crime.**

The conclusion was unavoidable: modernism's ideas themselves were seriously flawed. Critic Charles Jencks famously ascribed the "death of modernism" to the precise moment of 3.32 p.m. on July 15, 1972, when high-rise slab blocks in the notorious Pruitt- Igoe housing project in St. Louis, Missouri were ARCHITECTURE Urban Design for Architects | 7 professionally imploded by the city . Completed as recently as 1955, the buildings had been abandoned and vandalized by their erstwhile inhabitants to a degree that made them uninhabitable. Pruitt-Igoe became the most visible symbol of modernism's decline and fall but the seeds of doubt and discontent had already been planted as early as the 1950s in the polemics of a group of younger architects and urbanists known as Team X.

In contrast to the abstraction of city plans inspired by Le Corbusier, the work of younger architects, such as Aldo van Eyck, Giancarlo De Carlo, Peter and Alison Smithson, Shadrach Woods, Georges Candilis and Jaap Bakema – all of whom came to prominence in the 1950s through their association with Team X – sought to enrich modernism with a sense of humanism and social reality that the simplistic Four Function model lacked. Through the 1970s and into the 1980s, architects sought ways to enrich and transform the overly simplistic concepts of modernist urbanism.

Ultimately this was to lead to a rejection of most modernist thinking about cities, but in the meantime the power of the modernist view of the city, with its single-use zones divided by major highways, and new large buildings constructed as singular objects in open space, still held sway. Indeed, elements of that modernist vision of the city remain with us today in the second decade of the 21st century.

## 11.5 ARCHITECTURAL THEORY MEETS ZONING PRACTICE

Despite what we might think from reading the preceding sections, the modernist view of city design in America, with buildings and uses parceled out into separate functional zones and connected together by large roads, does not owe its longevity to the leadership of architects. Instead this paradigm has persisted due to the steadfast grip on city form exercised by planners.

By a somewhat bizarre historical confluence of ideas in the 1930s, avant-garde European ideas about architecture and cities, mostly with a socialist and utopian bent, intersected with pragmatic American planning concepts based on business efficiency, real estate, and commercial development. When the leading European architects and planners such as Walter Gropius and Ludwig Mies van der Rohe fled Hitler and the rise of fascism, they landed safely in America, but found that Americans were not much interested in utopia, especially with socialist overtones.

In effect, the transatlantic voyage stripped the political agenda from avant-garde European ideas. But what was left was still powerful, perhaps even more so: the Four Function model of separate zones for separate functions fitted neatly into America's evolving planning practice.

As we have noted, traditional American cities were more compact, connected, and walkable, and served efficiently by public transport, but changes to this historical pattern began out in California, in Modesto, in 1885. The white middle and upper classes in that community created regulations to restrict laundries, operated exclusively by Chinese families, to poorer parts of town, away from white residential areas. Ten years later, in Los Angeles, that city established separate zoning districts for residential and industrial areas. Partly, this was common sense; it was unhealthy to live next door to industry that might be spewing out toxic fumes. But then the citizens of LA went further – they banned all business uses from residential districts. As these ideas spread to other communities in subsequent decades, with each separate use partitioned into a separate city zone, the urban fabric of traditional American towns began to unravel. No longer could everything be mixed together as was the traditional norm. Now everything had to be sorted out into separate land areas for separate uses.

These changes were encouraged during the 1920s by the "Standard State Zoning Enabling Act" of 1924, whereby the U.S. Department of Commerce promulgated strictly use-based zoning codes across the country on the presumption that clarity and efficiency were good for business. These new zoning laws could have required new developments to follow traditional pattern of older historic American towns. But they didn't. In effect the new zoning laws made the traditional urbanism of Main Street illegal.

No longer could uses be mixed together and public spaces tightly defined. The pedestrian-scaled spaces between buildings were progressively dismantled and redesigned for cars.

To understand why these ideas took hold so firmly, we need to remember that as the nation picked itself up after the Great Depression in the 1930s, and then emerged on the winning side in World War II, there was very little interest in looking backwards at the hardships of history. The future beckoned, one that was cleaner, efficient and mechanized. Cities began to be thought of as giant machines, and "efficiency" became the most important concept. Efficiency became synonymous with simplicity -- and simplicity with single-use zoning.

If one was a homebuilder, building only one particular kind of housing, say, single-family detached homes, became the most simple, efficient, and profitable way to operate. Let someone else build apartments. Each developer thus became a specialist focused on a single product, be it different types of housing, office parks or shopping centres. Each type of development gained efficiency by simplifying its operation and excluding other kinds of buildings. Here, single-use zoning was a boon to private sector developers. Land could be divided up in advance for the different uses that – if you put them all together - would have made a town. But once they were legislated and built apart from each other, the traditional physical fabric of urban America was progressively dismantled.

**We started living our modern lives in separate compartments.**

This dissolution of America's traditional communities in the name of an optimistic, new, technologically based future was hastened by the development of one particular new technology: the motorcar. After WWII, cities were planned for the automobile. Different pieces of the city could now be spread widely apart; but it still took only a few minutes to go from one part to the other, so long as you travelled at 30 miles an hour, or more – which was easy on the new wide roads. To accommodate all these new cars, mid-twentieth century zoning codes required large parking lots for each use, covering the growing suburban landscape with tens of thousands of acres of asphalt. The zoning innovations of the 1920s became locked into proscriptive legislation by the 1950s and 1960s. Indeed, most municipal zoning codes in America today are still based on the approach set out in that model legislation established in 1924. These concepts hardened into dogma fifty years ago and have since been plastered with countless "band aid" amendments, trying to keep abreast of change. By the late 1980s most municipal zoning ordinances had become so confusing, and so devoid of anything resembling considerations of good urban design that a new wave of professionals took it upon themselves to institute radical changes. But most of these concerned professionals were not planners. They were architects, reclaiming the lost art of civic design.

## 11.6 THE REVIVAL OF TRADITIONAL URBANISM AND THE BIRTH OF "NEW URBANISM"

By the end of the 1980s, it became clear to members of the new generation of architects that what had been truths for their forbears had become anathema to their new appreciation of American cities. Many found they no longer believed what they had been taught.

Faced with this ideological void, groups of younger architects sought to construct a new set of beliefs, and many premises of modernist urbanism were overturned in this process. Many aspects of the search for new concepts focused around the recovery of more human scaled spaces and an architectural vocabulary that reconnected with public taste and urban history.

Specifically, urban design resurrected many of the precepts of traditional urbanism, and in particular it renewed a focus on the street and other clearly defined spatial types found in American cities from earlier periods. The street in particular - once identified in modernist thinking as a major cause of urban squalor - has been reclaimed as the primary form of public space in American communities.

This renewed appreciation of traditional urban forms was presaged by Jane Jacobs in her landmark book The Death and Life of American Cities. Her description of the vitality and life on the streets in her New York neighbourhood contrasted poignantly with the crime and grime of the urban wastelands produced by urban renewal (see Fig..6). However, professional architects and planners largely dismissed her stinging criticism of modernist planning; during the 1960s her advocacy for the importance of traditional streets and cohesive neighborhoods fell on deaf ears. But by the 1980s Jacobs' book had become a standard text, establishing a strong counter-narrative about city design, one that recognized again the importance of traditional city forms and spaces. Le Corbusier, once the hero of the modern city, soon became the archvillain of this revisionist history, with his revolutionary and draconian proposals for "The City of Tomorrow" identified as the source of everything bad about modernist urbanism.

Also during the late 1970s and 1980s, radical rethinking about urban design emerged from academia, initially from progressive teachers in schools of architecture, including Cornell and Yale. At Cornell, a new kind of urbanism was taught by the revered Anglo-American urbanist Colin Rowe, in conjunction with visiting professionals such as Michael Dennis and Steven Peterson. This approach to urban design focused much more on the context of cities and their history, seeking a deeper understanding of the relationship between existing urban places and new architectural projects. And, marking a major break from modernist dogma and its fetish with single, separated uses, this new approach welcomed a return to a mixed-up and layered urbanism.

**Fig. 6: Greenwich Village, New York. Although considerably changed from the 1950s and 60s when Jane Jacobs lived nearby, Greenwich Village still embodies lessons from traditional urban design, with connected streets lined by buildings that create defined spaces for human activities. Photo: David Walters**

Meanwhile at Yale, the renowned architectural historian Vincent Scully taught courses on the urban form and building types of American traditional towns and cities. Two of Scully's graduate students in the early 1970s, Andres Duany and Elizabeth Plater- Zyberk, found this material particularly fascinating. A few years after graduating, these two pioneers founded their ground-breaking urban design firm DPZ in 1980 and assumed leadership roles in the development of what became known as Neo-Traditional Development (TND).

Stimulated by this new awareness, they were able to understand many things that were wrong with modernist city planning; this critique mirrored many observations by Jane Jacobs, but was now articulated with an extra edge of practicality that presaged radical design action. This critique also led them to formulate a new approach to zoning -- writing and diagramming new rules for development that captured the spirit and essence of the rediscovered traditional urbanism. These new zoning regulations, known today as "Form based Codes" encapsulated many qualities of American urbanism that had been effectively outlawed by the draconian use-based zoning techniques. Traditional Neighbourhood Development was one precursor to New Urbanism; the other was Transit- Oriented development (TOD), also developed during the 1980s on the west coast by Peter Colthorp, Daniel Solomon, Douglas Kelbaugh and others. If a renewed appreciation of traditional American urbanism and the neighbourhood unit were the main highlights of Traditional Neighbourhood Development (TND), the emphasis of Transit-Oriented Development (TOD) was made equally clear in its title: it renewed the connection between urban form and public transportation that had atrophied decades earlier after the demise of the streetcar. TOD embodied many similar and complimentary ideas as its TND companion concerning traditional urban patterns, but it evolved specifically from the concept of the "Pedestrian Pocket." This was essentially a small town, or "urban village" organized primarily with the needs of the pedestrian in mind, like the pre-automobile suburbs that formed the basis for TNDs, but developed around new public transit -- usually light rail – that enabled residents of one "pocket" to travel conveniently to others and to a major metropolis. Once again (as with Clarence Perry's 1927 Neighborhood Unit plan) the concept of the five-minute walk defined the scale of the development, five minutes being established today as the maximum distance an average American will walk to catch transit.

The architects behind the TOD movement added one other ingredient to the mix of recovered "new" ideas – energy efficiency. Several California-based designers had been involved with solar and passive energy designs in the 1970s, and this emerging interest in energy performance planted the seed of today's increasingly urgent focus on matters of sustainability and resilience.

Common to both converging design movements were parallel developments imported from Europe, from the work of Aldo Rossi in Italy and especially from Leon and Rob Krier, based out of Luxembourg. Leon Krier was especially influential - in the UK as Prince Charles' favorite architect - but more importantly in the USA as a theoretician who provided guidance to Duany and Plater-Zyberk's evolving urban design language. Krier's influence lent an increasingly neoclassical and historicist bent to the aesthetics of neo-traditional development. This was accepted by some urban designers as an effective way to reconnect with public taste and to root new development firmly in the Western tradition, but rejected by others who saw this as needless nostalgia. Even worse, it undercut the forward-looking agenda that accompanied the reconnection to history. To this more progressive group, the most important lessons from traditional urbanism were in the scale and formation of human scaled urban space and urban infrastructure, not in the aesthetics of buildings.

The confluence of Traditional Neighbourhood Development and Transit-Oriented Development led to the formalization in 1996 of the "New Urbanist" movement. The name "The New Urbanism" was consciously chosen to define the return to America's traditional urban forms and spaces from the period 1890-1930. It was defined as "new" in contrast to the old and discredited urban language of modernism.

And it was to be "urban" by creating a coherent urban structure to counteract the faults of a sprawling suburban model of city development. The movement's manifesto was written out at length in the Charter of The New Urbanism. It set forth a series of principles "to guide public policy, development practice, urban planning and [urban] design".

It was also no accident that this new Charter for the post-modern age was ratified at the 4th Congress of the new movement, in Charleston, S.C. in 1996. The declaration and signing of the Charter of the New Urbanism at the 4th Congress can be read as a deliberate repudiation and overwriting of the modernist concepts in the Charter of Athens, originally produced, as we have noted earlier, at a different 4th Congress, that of CIAM in 1933. The new Charter thus became a rallying cry for the redesign of American towns and cities.

By the end of the 20th Century, New Urbanism had matured into a detailed and multi-faceted approach to rebuilding America's towns and cities, with an increasingly long list of successful projects (see Fig. 7). As the 21st century has progressed to the time of writing. New Urbanism has evolved further to include an environmental agenda around concepts of sustainability and resilience, now vital urban design issues in the face of climate change and ecologically damaging suburban sprawl.

**Fig. 7: Birkdale Village, Huntersville, North Carolina, 1999-2003. Shook Kelley, architects. Crosland and Pappas Properties, developers. The 52-acre mixed-use centre is linked to adjacent housing through a network of connected streets and parks. The urban form of the development is directly attributable to Huntersville's form-based zoning ordinance, written in 1996 by David Walters and Ann Hammond.**

# RELATED TERMINOLOGY

**Accessibility:** The ease of reaching destinations. In a highly accessible location, a person, regardless of age, ability or income, can reach many activities or destinations quickly, whereas people in places with low accessibility can reach fewer places in the same amount of time. The accessibility of an area can be a measure of travel speed and travel distance to the number of places ('destination opportunities') to be reached. The measure may also include factors for travel cost, route safety and topography gradient.

**Active Frontage:** Refers to street frontages where there is an active visual engagement between those in the street and those on the ground and upper floors of buildings. This quality is assisted where the front facade of buildings, including the main entrance, faces and opens towards the street. Ground floors may accommodate uses such as cafes, shops or restaurants. However, for a frontage to be active, it does not necessarily need to be a retail use, nor have continuous windows. A building's upper floor windows and balconies may also contribute to the level of active frontage. Active frontages can provide informal surveillance opportunities and often improve the vitality and safety of an area. The measures of active frontage may be graded from high to low activity.

**Active Use:** Active uses are uses that generate many visits, in particular pedestrian visits, over an extended period of the day. Active uses may be shops, cafes, and other social uses. Higher density residential and office uses also can be active uses for particular periods of the day.

**Activity Centre:** Activity centres within cities and towns are a focus for enterprises, services, shopping, employment and social interaction. They are where people meet, relax, work and often live. Usually well-served by public transport, they range in size and intensity of use from local neighbourhood strip shopping centres to traditional town centres and major regional centres. An activity centre generally has higher intensity uses at its central core with smaller street blocks and a higher density of streets and lots. The structure of activity centres should allow for higher intensity development, street frontage exposure for display and pedestrian access to facilities.

**Adaptability (or 'adaptive re-use'):** The capacity of a building or space to respond to changing social, technological, economic and market conditions and accommodate new or changed uses.

**Amenity:** Urban amenities means urban facilities such as parks, playgrounds, green spaces, parking facilities, public wi-fi facilities, public bus transport, bus shelters, taxi and rickshaw stands, libraries, affordable hospitals, cultural centres, recreation centres, stadium, sports complex and any other urban facility that the State Government may, on the recommendation of the Authority, specify to be an urban amenity, but does not include infrastructure development work.

**Arcology:** What happens when you splice the words "Architecture" and "Ecology." Used to describe self-contained megastructures that reduce human impacts on the environment (basically, the conceptual projects that architects love to design and no-one loves to pay for.)

**Arterial Road:** The principal routes for the movement of people and goods within a road network. They connect major regions, centres of population, major transport terminals and provide principal links across and around cities. Arterial roads are divided into primary and secondary arterial roads. Declared arterial roads are managed by VicRoads. Also see 'Major roads'.

**Articulation:** "Articulation" means to make something — a process, a function, or a form — readable, so people can interact with it through their perceptions. Previous literature has linked urban form to the function it enables, the way of life that its function connotes, and the overall ambience that people feel.

**Average Building Volume Per Inhabitant:** Variable according to the intended use, as a rule, for residential areas, it varies between 100-150 cubic meters / inhabitant. Land use index: Defines the maximum gross floor area (S.L.P.) achievable for each unit of land area.

**Background Buildings:** Buildings that lack individual architectural merit but contribute to the overall character of an area or district; simple commercial buildings in a historic district function as background buildings.

**Barriers and Fences:** Barriers such as bollards and fences can define boundaries and protect people from traffic hazards and level changes. They also protect trees and shrubs from people and vehicles. A barrier may be made as bollards, screens, rails, fences, kerbs and walls. Barriers and fences can provide an opportunity for public art or to communicate local stories. They may also provide opportunities for seating.

**Blank Wall:** A wall which has few or no windows or doors, and has no decoration or visual interest. See also active frontage.

**Blue Space:** Blue space is an urban design term for visible water. Attractive blue spaces such as waterfront parks, harbors, ports, marinas, rivers, open air streams, canals, lakes, ponds and fountains are thought to improve quality of life and help to moderate urban heat islands. Many cities have highly industrialized waterfront zones that are built out with artificial land. Other cities have disrupted access to the sea with poorly designed seawalls and other barriers.

**Brownfield:** Brownfield is a term for land that has been previously used for industrial or commercial purposes that is polluted or feared to be polluted. Such sites often require an expensive clean up and it is common for governments to require the parties responsible to pay such costs.

**Building Cap:** Maximum allowable construction in a designated area or city; for example, San Francisco limits annual down town office space construction to 475,000 square feet and Petaluma, CA, limits the number of residential building permits issued annually.

**Building Line:** A line usually set with respect to the frontage of a plot of land which is fixed by statute or by deed or contract and beyond which the owner of the land may not build.

**Building Volume:** It is understood as the product of the sum of the GFA of the individual floors of the building for the virtual height, conventionally equal to 3 m. It is a value that may vary from municipality to municipality (for example the R.E. of Milan provides a virtual height of 3 m, in Rome instead of 3.20 m), regardless of its actual height. The maximum volume that can be built in a building sector is also deducted on the basis of the building density indices allowed by the P.R.G. for that area, i.e. by multiplying the St or Sf respectively by It or If.

**Buildings in Activity Centres**: Buildings in activity centres accommodate a wide range of uses, such as living, working, shopping and services. Buildings in these locations may be larger than those in surrounding neighbourhoods, occupy more of the site area and be built to the front and side boundaries. They may incorporate a mix of uses that mean people are present at different times of the day.

**Build-to-Line:** A zoning device which controls the location of buildings to create consistent streetwalls or define public spaces; unlike a set back, which establishes a minimum distance from a property line or street, a build-to-line establishes the maximum permitted set back or exact location of a building facade

**Built Form:** Built form refers to the function, shape and configuration of buildings as well as their relationship to streets and open spaces. Built form refers to the function, shape and configuration of buildings as well as their relationship to streets and open spaces.

**Character:** Character areas are an important tool in helping to deliver contextually responsive urban design, allowing the urban designer to understand and respond to the unique qualifies of any parficular site or neighbourhood.

**Circulation Space:** Circulation spaces are part of the common area of a commercial, mixed use or higher density residential building and are used by occupants, residents and other building users. These spaces include foyers, corridors, car parking areas, and garden and recreation areas.

**City:** A city is a large human settlement. It can be defined as a permanent and densely settled place with administratively defined boundaries whose members work primarily on non-agricultural tasks

**Community:** A community is social group that are bound together by geography, profession, lifestyle, circumstances, believe or interests.

**Connectivity:** The number of connecting routes within a particular area, often measured by counting the number of intersection equivalents per unit of area. An area may be measured for its 'connectivity' for different travel modes – vehicle, cyclist or pedestrian. An area with high connectivity has an open street network that provides multiple routes to and from destinations.

**Coverage Ratio:** It defines the maximum amount of covered area (Sc) in relation to the land area of the lot (Sf). Rc = Sc / Sf It basically determines how much of the lot can be occupied by buildings and how much left free as uncovered space.

**Covered Area:** Covered area means the horizontal projection of the parts built above ground, with the exclusion of: • projecting bodies (balconies, eaves, cornices, etc. with overhang not exceeding 1.20 m or as specified by local legislation; • canopies covering the entrances (if less than 8-10 square meters of surface); • parts of the building completely underground; • outdoor pools and tanks, threshing floors, fertilizers and cultivation greenhouses.

**Cul-de-sac:** A street with only one inlet/outlet connected to the wider street network. A closed cul-de-sac provides no possible passage except through the single road entry. An open cul-de-sac allows cyclists, pedestrians or other non- automotive traffic to pass through connecting paths at the cul-de-sac head.

**Culture:** Culture are the aspects of life that people value and enjoy. It is considered a defining characteristic of humanity that includes things like language, art, music, architecture, customs, rituals, pastimes, festivals, cuisine, fashion, history, stories and myth. The following are elements and variants of culture.

**Demolition:** Demolition is the tearing-down of buildings which involves taking a building apart while preserving the valuable elements for re-use. There are various methods of demolition. The building is brought down either manually or mechanically depending upon the method used for demolition of buildings

**Density:** Urban density is a term used in urban planning and urban design to refer to the number of people inhabiting a given urbanized area.

**Design Response:** Explanation and demonstration of how a proposed building development or public space design is informed by and responds to the site and context analysis.

**Desire-line (or 'pedestrian desire-line'):** The desire-line path usually represents the preferred route and the shortest or most easily navigated route between an origin and destination. Desire- lines can often be seen as alternative shortcut tracks in places where constructed pathways take a circuitous route. They are almost always the most direct and the shortest route between two points.

**Development:** Urban development covers infrastructure for education, health, justice, solid waste, markets, street pavements and cultural heritage protection. These constructions usually form part of specific sector programmes, including capacity building measures.

**Efficiency:** Efficiency is the output of something in comparison to its maximum potential. It is the opposite of waste. Calculating efficiency requires a measurement of maximum potential. As such, it is used to describe highly measurable processes and machines. The work of people is described by a similar term, productivity. The following are common business efficiency terms.

**Enclosure:** Enclosure refers to the extent to which buildings, walls, trees and other vertical items frame a street and public space. Public spaces that are framed by vertical elements in relative proportion to the width of the space between the elements have a room-like quality that is comfortable for people

**Encroachment:** Or Urban Encroachment or Urban sprawl is defined as "the spreading of urban developments on undeveloped land near a city". Urban sprawl has been described as the unrestricted growth in many urban areas of housing, commercial development, and roads over large expanses of land, with little concern for urban planning.

**Fabric or Urban Fabric:** The term 'urban fabric' describes the physical characteristics of urban areas, that is, cities, and towns. This includes the streetscapes, buildings, soft and hard landscaping, signage, lighting, roads and other infrastructure. Urban fabric can be thought of as the physical texture of an urban area

**Facade:** The principal wall of a building that is usually facing the street and visible from the public realm. It is the face of the building and helps inform passers-by about the building and the activities within.

**Facadism:** A practice vehemently hated by many architects, it mostly consists of badly hiding a glass box behind a skinned heritage building.

**FAR (floor area ratio):** A formula for determining permitted building volume as a multiple of the area of the lot; the FAR is determined by dividing the gross floor area of buildings on a lot by the area of the lot; for example, a FAR of 6 on a 5,000 square foot lot would allow a building with gross area of 30,000 square feet

**Filtering Surface (sqm):** By filtering surface we mean the one arranged in green, not built either at the level of the pavement or underground.

**Floor Area Ratio:** Total floor area of building. Area of the plot.

Form Follows Nature: Form follows nature is an architecture and design technique that uses shapes and forms found in nature such as plants, animals, insects, geological and astronomical shapes. Such techniques may achieve highly functional designs or may be purely aesthetic as a means of integrating with natural surroundings. Form follows nature is a central principle of organic architecture along with other principles of sustainable design and natural aesthetics.

**Fused Grid:** A type of street network pattern that looks like an IQ test.

**Gross Floor Area:** Gross floor area means the sum of the gross floor area of each floor of the building measured within the external profile of the perimeter walls of the various floors and intermediate floors, both above ground and underground. The surfaces used for the shelter of cars, with the relative spaces for maneuvering and access, cellars, open overhangs (terraces, balconies, loggias), arcades, non-habitable attics and the technical volumes of the building are normally excluded from the calculation. . In the case of basement floors, the areas used for laboratories, offices, warehouses and meeting rooms must be taken into account (always see local regulations).

**Height of a Building:** is the maximum height between those of the various fronts of the building itself, measured: • from the floor of use in front of the front up to the intrados of the last floor, for buildings with flat roofs or with slopes up to 35%; • from the usage plan to the midpoint of the roof structure if this has slopes greater than 35%; • in the case of buildings located along sloping land, the height is normally measured at the midpoint of the front.

**Land Area:** By land area we mean the surface of the building lot net of surfaces for primary and secondary urbanization works. In the land area, the areas to be used for construction, pedestrian paths and possibly private parking areas are identified.

**Land Building Index:** It is the ratio (m3 / m2) between the achievable volume and the area to be built, excluding road sites, even if private or to be sold to the Municipality. It then defines the maximum building volume on each unit of land area.

**Land Density Index:** It defines the maximum number of theoretical settlers for each unit of land area expressed in hectares. Minimum distance of buildings from the property border: It is determined by measuring the distance between the projection of the external surfaces of the perimeter walls of the building, net of the open projecting bodies, and the property boundary at the closest point of the building itself.

**Land Use Index:** Defines the maximum gross floor area (S.L.P.) achievable for each unit of territorial surface.

**Maximum Height of Buildings:** Defines the maximum real height allowed. It is measured from the average height of the sidewalk along the main front of the building, to the maximum height of the intrados (unless otherwise prescribed by the building regulations indicating the extrados) of the roof slab of the last habitable floor.

**Minimum distance Between Buildings:** It is determined by measuring the minimum distance between the walls net of the open projecting bodies (Ministerial Decree no. 1444/1968). The municipal regulations specify the distances, in the various homogeneous areas, when different from the national standard.

**Occupied Area (sqm):** Occupied area means the covered area increased by any basement quota - including the ditches -, exceeding the covered area, and the projection of arcades, even if not calculated for the purposes of the covered area. The areas destined to lanes and vehicle ramps, surface parking lots and consolidated pedestrian paths contribute to form the occupied surface.

**Parking Area:** This refers to the area to be allocated to garages or covered parking spaces pertaining to the housing organization, including maneuvering spaces. The parking area (Sp) must be contained within 45% of the usable living area. The 45% limit is not intended for single accommodation but refers to the total useful surface (Su) of the housing organization. The aforementioned percentage may be waived in the presence of housing structures consisting mainly of dwellings with a useful living area (Su) of less than 60 square meters.

**Permeability Ratio:** It defines the minimum quantity of filtering surface, that is the minimum quantity of the lot surface to be maintained or green with the exclusion of any building, including underground, or paving, expressed as a percentage of the Sf.

**Permeability:** How cheese hole-y an urban area is. New Urbanists love this.

**Placemaking:** The art of making "places" rather than stand-alone pretty buildings.

**Protected View:** When a view is so beautiful you have to protect it.

**Territorial Density Index:** It defines the maximum number of theoretical settlers for each unit of territorial surface expressed in hectares (1 ha = 10,000 m2).

**Territorial Manufacturability Index:** It defines the maximum building volume on each unit of land area, excluding the volume relating to urbanization works.

**The Buildable Volume Includes:** • the above ground part of existing buildings and / or to be built on the lot; • the underground part of the same buildings, if intended for residence, offices or production activities; • ancillary buildings, for their part above ground.

**Usable Area of the Accommodation:** The useful area of the accommodation means the floor area measured net of the perimeter and internal walls, pillars, doorways and French doors, ventilation or flue pipes, any chimneys, shafts, stairs uncommon internal and lodges.

**Virtual H (ml):** It is the conventional height value, which multiplied by the GFA, quantifies the volume of a building regardless of its actual height.

# REFERENCES

AbdulgaderA.AinaY. A.2005Sustainable cities: An integrated approach to sustainable urban design. In: Sustainable Development and Planning II: 1A. Kungolos, C. A. Brebbia & E. Beriatos, (Ed.), WIT Press, UK, 1524.

Abukhater, A. & Walker, D. (2010). Making smart growth smarter with GeoDesign. Directions Magazine, Available from http://www.directionsmag.com/articles/making-smart-growth-smarter-with-geodesign/122336

Adhya A. Plowright P. Stevens J.2 010Defining sustainable urbanism: Towards a responsive urban design. In: Proceedings of the Conference on Sustainability and the Built Environment, 24King Saud University, Riyadh, Saudi Arabia

Adler, J. (May 14, 1995). *Bye-Bye Suburban Dream: 15 Ways to Fix the Suburbs.* Newsweek.

AdolpheL.2001A simplified model of urban morphology: Application to an analysis of the environmental performance of cities. Environment and Planning B: Planning and Design, 28, 183200.

Alcorn, J. B. (1981). Huastec Noncrop Resource Management: Implications for Prehistoric Rain Forest Management. *Human Ecology, 9*(4), 395-417.

Al-hathloulS. A.1981Tradition, continuity and change in the physical environment: the Arab-Muslim city.PhD Thesis, Massachusetts Institute of Technology, USA.

Al-hemaidiW. K.2001The metamorphosis of the urban fabric in Arab-Muslim city: Riyadh, Saudi ArabiaJournal of Housing and the Built Environment162179201.

Allard F, Ghiaus C, Szucs A. Naturai Ventilation in High-Density Cities. In: Ng E, editor. Designing High Density Cities - For Social and Environmental Sustainability. London, UK: Earthscan Publications Ltd.; 2009. p. 137-62.

Allard F, Ghiaus C. Natural Ventilation in the Urban Environment: Assessment and Design. London, UK: Routledge; 2005. 266 p.

Allen, A. (April 1, 2003). Environmental planning and management of the peri-urban interface: perspectives on an emerging field. [Journal]. *15*, 15.

Anthopoulos, Leonidas G. (2015). Understanding the Smart City Domain: A Literature Review. Springer International Publishing Switzerland. pp. 9-18.

Batty, M.; Axhausen, K. W.; Giannotti, F.; Pozdnoukhov, A.; Bazzani, A.; Wachowicz, M. (2012). Smart Cities of the Future, *The European Physical Journal Special Topics*, Vol.214, No.1, pp. 481–518.

BattyM.DodgeM.JiangB.SmithA.1998GIS and urban designWorking Paper Series, 3, CASA, University College London, UK

BiancaS.2000Urban form in the Arab world: Past and present.Thames and Hudson Ltd, London, UK.

BMT WBM Pty Ltd. (2009). *Evaluating options for water sensitive urban design: A national guide.* Joint Steering Committee for Water Sensitive Cities (JSCWSC), Australia.

Broto, V.C. (2017). Urban Governance and the Politics of Climate Change. World Development. 93:1–15.

Brower, S. (2002). 'The Sectors of the Transect.' *Journal of Urban Design*, Vol. 7, No. 3, October, 313-20.

Bullis, K. (2009). A zero-emissions city in the desert. *MIT Technology Review* 56–63.

BURNS, M. The alcohol problem in Los Angeles. *Abstracts*, 4: 9–15 (1983).

Butti K, Perlin J. Solar Architecture in Ancient Greece. A Golden Thread: 2500 Years of Solar Architecture and Technology. New York, USA: Van Nostrand Reinhold; 1980. p. 289.

Caltabiano I. Consapevolezza energetica nelle costruzioni tradizionali in area mediterranea. Ingegno e natura al servizio dell'abitare. In: Biondi B, editor. 1st International Research Seminar on Architectural Heritage and sustainable development on small and medium cities in south mediterranean regions Results and strategies of research and cooperation. Pisa, Italy: Edizioni ETS; 2005. p. 470-8.

Campbell, S. (Summer 1996). Green Cities, Growing Cities, Just Cities? Urban Planning and the Contradictions of Sustainable Development. *Journal of the American Planning Association*, 30.

CAPRA, F. *The turning point: science, society and rising culture.* New York, Simon and Schuster,1982.

CarmonaM.2001Sustainable urban design- A possible agenda, In: Planning for sustainable future, A. Layard, S. Davoudi & S. Batty, (Ed.), Spon Press, London, UK.

CASSEL, J. The contribution of the social environment to host resistance. *American journal of epidemiology*, 104: 107–123 (1976).

CATALANO, R. *Introduction to public health.* Lecture presented in Public Health 200 1997. School of Public Health, University of California, Berkeley, CA, 1997.

ChoguillC. L.2008Developing sustainable neighbourhoodsHabitat International3214148

Chow, W. T., Salamanca, F., Georgescu, M., Mahalov, A., Milne, J. M., and Ruddell, B. L. (2014). A multi-method and multi-scale approach for estimating city-wide anthropogenic heat fluxes. *Atmospheric Environment* 99, 64–76.

*City health planning: the framework.* Copenhagen, WHO Healthy Cities Project Office, 1995.

Clarion Associates. (2008). Abu Dhabi estidama program: Interim estidama community guidelines assessment system for commercial, residential, and institutional development. Prepared for the Emirate of Abu Dhabi Urban Planning Council.

Clay, J. (1981) Romanticism, New York, NY: Vendome Press. Cleveland City Planning Commission (1974) Cleveland Policy Planning Report.

Climate change in Hong Kong. Technical Note No. 107, Hong Kong Observatory.

COHEN, L. A public health approach to the violence epidemic in the United States. *Environment andurbanization*, 5: 50–66 (1993).

Coleman, R. (1978). *Attitudes towards Neighborhoods: How Americans Choose to Live.* Working Paper No. 49, Cambridge, Massachusetts: Joint Center for Urban Studies of MIT and Harvard University.

Collaboration between architects and planners in an urban design studio 21 Wender, W. and Roger, J. (1995) 'The design life space: verbal communication in the architectural design studio', Journal of Architectural and Planning Research, Vol. 12, No. 4, pp.319-336.

Communities and Local Government Committee, House of Commons. (2008). *New Towns: Followup. Ninth Report of Session 2007-2008.* London: The Stationery Office.

*Community participation in local health and sustainable development: a working document onapproaches and techniques* Copenhagen, WHO Regional Office for Europe, 1999 (European Sustainable Development and Health Series, No. 4).

Curl, J. (1999) A Dictionary of Architecture, Oxford, UK: Oxford University Press.

Dalton, L. (2001) 'Weaving the fabric of planning as education', Journal of Planning Education and Research, Vol. 20, pp.423-436.

Dameri, R. P. (2017). The Conceptual Idea of Smart City: University, Industry, and Government Vision, in Smart City Implementation. Springer Link. pp. 23–43.

Dangermond, J. (2009). GIS, design, and evolving technology. ArcNews, Fall 2009, Available from http://www.esri.com/news/arcnews/fall09articles/tofc-fall09.html

Decker, E., Smith, B., and Rowland, F. (2000). Energy and material flow through the urban ecosystem. *Annual Review of Energy and the Environment* 25, 685–740.

Dell'Osso R. L'architettura della villa. Maggioli Editore; 2008. 185 p.

DETR. (2000). *Our Towns and Cities: the Future: Delivering an Urban Renaissance.* London: The Stationery Office.

Dewey, J. (1902/1990) The Child and the Curriculum, Chicago, IL: University of Chicago.

Dewey, J. (1916) Democracy and Education: An Introduction to the Philosophy of Education, New York, NY: Macmillan.

Dewey, J. (1933) How We Think, a Restatement of the Relation of Reflective Thinking to the Educative Process, London, UK: Harrap.

Dewey, J. (1938) Experience and Education, New York, NY: Macmillan.

Dickie, S., Ions, L., McKay, G., and Shaffer, P. (2010). *Planning for SuDS – Making it Happen.* Ciria. Donzelot, J. (2009). *La ville à trois vitesses*, Éditions de la Villette, 2009.

Doulos, L., Santamouris, M., and Livada, I. (2004). Passive cooling of outdoor urban spaces. The role of materials. *Solar Energy* 77(2), 231–249.

Ellin, N. (1999) Postmodern Urbanism, New York, NY: Princeton Architectural Press.

Erell, E., Pearlmutter, D., and Williamson, T. (2012). *Urban Microclimate: Designing the Spaces Between Buildings.* Routledge.

Falkner, Nick. (2018). 2018 Smart City Snapshot. University of Adelaide. https://www.lga.sa. gov.au/webdata/resources/minutesAgendas/Smart%20City%20Slides%20Jan%2019%20for%20MLGG.pdf

FathyH.1973Architecture of the poor. University of Chicago Press, Chicago, USA

Faust BC. Andeutungen über das Bauen der Häuser und Städte zur Sonne. Hannover, Germany: Hahn'sche Hofbuchhandlung; 1829.

FINGERHUT, L.A. & KLEINMAN, J.C. International and interstate comparisons of homicide among young males. *Journal of the American Medical Association*, 263: 2210–2211 (1990).

Fisher, T. (2000) In the Scheme of Things: Alternative Thinking on the Practice of Architecture, Minneapolis, MN: University of Minnesota Press.

Flörke, M., Wimmer, F., Laaser, C., Vidaurre, F., Tröltzsch, J., Dworak, T., Stein, U., Marinova, N., Jaspers, F., Ludwig, F., Swart, R., Long, H., Giupponi, G., Bosello, F., and Mysiak, J. (2011). *Final Report for the Project Climate Adaptation – Modeling Water Scenarios and Sectoral Impacts.* CESR – Center for Environmental Systems Research.

Gans, H. (1968) 'From urbanism to policy-making', Journal of the American Institute of Planners, Vol. 36, No. 4, pp.223-225.

Garcia, R. (1993) Changing Paradigms of Professional Practice, Education and Research in Academe: A History of Planning Education in the United States, Doctoral Dissertation available from the University of Pennsylvania, Philadelphia, PA.

Garnier T. Una Città Industriale. Milano, Italy: Jaca Book; 1990.

Gehl, J. (2011). *Life Between Buildings: Using Public Space*. Washington, D.C. Island Press.

Georgescu, M., Morefield, P. E., Bierwagen, B. G., and Weaver, C. P. (2014).

GHEORGHE BREAZU, C. D. Fire Risks in the Field of Architecture and Urban Planning Design Process of the Civil Constructions, Management, Evaluation and Control. 16.

Gideon S. Spazio, Tempo Architettura. 2nd ed. Milano, Italy: Hoepli; 1987.

Givoni B. Urban design in different climates. World Meteorol Organ WMO/TD. 1989;(346).

GLASGOW, C. Local action – global solution, a strategy for a sustainable city. http:// www.archinet. co.uk/andromeda/top.html (retrieved in April 1999).

Gold, B. (1985). Foundations of Strategic Planning for Productivity Improvement. *Interfaces, 15*(3), 15-30.

Gold, J. R. (1997). *The Experience of Modernism: Modern architects and the future city.* London. E & FN Spon.

Gomez-Ibanez, D. J., Boarnet, M. G., Brake, D. R., Cervero, R. B., Cotugno, A., Downs, A., Hanson, S., Kockelman, K. M., Mokhtarian, P. L., Pendall, R. J., Santini, D. J., and Southworth, F. (2009). *Driving and the built environment: The effects of compact development on motorized travel, energy use, and CO2 emissions.* Oak Ridge National Laboratory (ORNL).

Goodchild M. F.2010Towards GeoDesign: Repurposing cartography and GIS? Cartographic Perspectives, 66, 5569

Grimmond, S. (2007). Urbanization and global environmental change: Local effects of urban warming. *Geography Journal* 173, 83–88.

Gupta V. Thermal efficiency of building clusters: an index for nonair-conditioned buildings in hot climates. In: Hawkes D, editor. Energy and Urban Built Form. Oxford, UK: Butterworths, UK; 1987.

Guttikunda, S. K., Carmichael, G. R., Calori, G., Eck, C., and Woo, J. H. (2003). The contribution of megacities to regional sulfur pollution in Asia. *Atmospheric Environment* 37(1), 11–22.

Guzetta, J. and Bolens, S. (2003) 'Urban planners' skills and competencies', *Journal of Planning Education and Research,* Vol. 23, pp.96-106.

Hachema C, Athienitisb A, Fazio P. Evaluation of energy supply and demand in solar neighborhood. Energy Build. 2012; Volume 49:335-47.

Hajduk, Sławomira. (2016). The Concept of a Smart City in Urban Management, *Business, Management and Education* Vol. 14, No.1, pp. 34–49. DOI:10.3846/bme.2016.319.

HakimB.1988Recycling the experience of traditional Islamic urbanism. In: Preservation of Islamic architecture heritage, Arab Urban Development Institute, Riyadh.

HALL, P. *Cities of tomorrow*. Malden, MA, and Oxford, Blackwell Publishers, 1996.

Hall, R. (2006). *Understanding and Applying the Concept of Sustainable Development to Transportation Planning and Decision-Making in the U.S.* Doctoral dissertation, MIT.

HANCOCK, T. & DUHL, L. *Promoting health in the urban context*. Copenhagen, WHO RegionalOffice for Europe, 1988 (WHO Healthy Cities papers, No. 1).

Harzallah A, Siret D, Monin E, Bouyer J. Controverses autour de l'axe héliothermique: L'apport de la simulation physique à l'analyse des théories urbaines. INHA [Internet]. 2014; Available from: http://inha.revues.org/2509

Harzallah A. Émergence et évolution des préconisations solaires dans les théories architecturales et urbaines en France, de la seconde moitié du XIXe siècle à la deuxième guerre mondiale. University of Nantes; 2007.

Hatch, C. (1984) The Scope of Social Architecture, New York, NY: Van Nostrand Reinhold.

Healey, P. (1997) Collaborative Planning: Shaping Places in Fragmented Societies, Vancouver, Canada: University of British Columbia Press.

Heiligenthal RF. Deutsche Städtebau. 1st ed. Heidelberg, Germany: Carl Winter; 1921.

Herder P., Turk, A., Subrahmanian, E. and Westerberg, A. (2003) 'Communication and collaborative learning in cross-Atlantic design course', Journal of Design Research, Vol. 3, No. 2.

Hilberseimer L. Großstadtarchitektur. L'Archtiettura della Grande Città. Napoli, Italy: Clean; 1998.

Hirt, S. (2005) 'Toward postmodern urbanism: Evolution of planning in Cleveland, Ohio', Journal of Planning Education and Research, Vol. 25, No. 1, pp.27-42.

Hoyer, J., Dickhaut, W., Kronawitter, L., and Weber, B. (2011). *Water sensitive urban design*. Jovis Verlag GmbH.

Inam, A. (2002) 'Meaningful urban design: teleological/catalytic/relevant', Journal of Urban Design, Vol. 7, No. 1, pp.35-58.

Innes, J. (1996) 'Planning through consensus-building', Journal of the American Planning Association, Vol. 62, No. 4.

JabareenY. R.2006Sustainable urban forms their typologies, models, and concepts. Journal of Planning Education and Research, 2613852

Jacobs, J. (1961) The Death and Life of Great American Cities, New York, NY: Random House.

Jacobs, J. (1961) *The Death and Life of Great American Cities*. New York: Random House.

Jacobs, J. (1961). *The Death and Life of Great American Cities*. New York: Random House.

JACOBS, J. *The death and life of great American cities*. New York, Random House, 1961.

Jacobson, M. Z., and Ten Hoeve, J. E. (2012). Effects of urban surfaces and white roofs on global and regional climate. *Journal of Climate* 25(3), 1028–1044.

Jencks, C. (1977). *The Language of Postmodern Architecture*. London: Academy Editions.

JenksM.BurgessR.2000Compact cities: Sustainable urban forms for developing countries. Spon Press, London, UK.

Jiao, Y. (2011, 17-19 June 2011). *Coordinated urban-rural development and the transition of planning management &#x2014; A case study of Chengdu, China.* Paper presented at the China Planning Conference (IACP), 2011 5th International Association for.

Johnson, D. and Johnson, R. (1994) Learning Together and Alone, Englewood Cliffs, NJ: Prentice-Hall.

Jon, B. (1978). Toward an Applied Human Ecology for Landscape Architecture and Regional Planning. *Human Ecology, 6*(2), 179-199.

Jonassen, D. (1991) 'Objectivism vs constructivism: do we need a new philosophical paradigm?', Educational Technology Research and Development, Vol. 39, pp.5-14.

Jones, C., and Kammen, D. M. (2014). Spatial distribution of U.S. household carbon footprints reveals suburbanization undermines greenhouse gas benefits of urban population density. *Environmental Science and Technology* 48(2), 895–902.

KANES-WEISMAN, L. *Discrimination by design: a feminist critique of the man-made environment.*Urbana, IL, University of Illinois, 1992.

Keane, T., and Walters, D. (1995). *The Davidson Land Plan*, Davidson, NC: Town of Davidson.

Kelbaugh, D. (2004) 'Seven fallacies in architectural culture', Journal of Architectural Education, Vol. 58, No. 1, pp.66-68.

Kelbaugh, D. editor, 1989. *The Pedestrian Pocket Book*.

Kelly, E., Becker, B. and So, F. (1999) Community Planning: An Introduction to Comprehensive Planning, Washington DC: Inland Press.

Kemp, R.L. and Stephani, C.J. *Cities Going Green: a handbook of best practices*. Jefferson, N.C. McFarland & Company Inc. pp. 117-120.

Khakee, A., & Strömberg, K. (1993). Applying Futures Studies and the Strategic Choice Approach in Urban Planning. *The Journal of the Operational Research Society, 44*(3), 213-224.

KhanS. M.1981The influence of Arabian tradition on the old city of Jeddah: The urban setting, In: The Arab city: Its character and Islamic cultural heritage, I. Serageldin & S. El-Sadek, (Ed.), Arab Urban Development Institute, Riyadh, Saudi Arabia.

KHONG, J. The community park user's survey: a partnership approach to the planning process.

Klein, J. (2005) Humanities, Culture and Interdisciplinarity: The Changing American Academy, Albany, NY: State University of New York Press.

Knowles RL. Energy and Form: Ecological Approach to Urban Growth. MIT Press; 1974. 198 p.

Knowles RL. Sun Rhythm Form. Cambridge, Mass: The MIT Press; 1985. 304 p.

Knowles RL. The solar envelope: Its meaning for energy and buildings. Energy Build. 2003;35(1):15-25.

Knowles, R. (2003). The solar envelope: Its meaning for energy and buildings. *Energy and Buildings* 35, 15–25.

Knox, P. (1988) Urbanization: An Introduction to Urban Geography, Englewood Cliffs, NJ: Prentice Hall.

Koch, A., Schwennsen, K., Dutton, T. and Smith, D. (2002) The Redesign of Studio Culture: A Report of AIAS Studio Culture Task Force, Washington, DC: American Institute of Architecture Students.

Kockelmans, J. (Ed.) (1979) Interdisciplinarity and Higher Education, University Park, PA: Pennsylvania State University Press.

Komninos, Nicos; Mora, Luca. (2018). Exploring the Big Picture of Smart City Research, *Scienze Regionali*. Vol.17, No.1, pp. 1-17.

Konrad, C. P. (2003). Effects of urban development on floods. U.S. Geological Survey Fact Sheet 076–03.

Kratzer FA. Das Stadtklima. 2nd ed. Braunschweig: Vieweg; 1956.

Kravčík, M., Pokorný, J., Kohutiar, J., Kovác☐, M., and Tóth, E. (2007). *Water for the Recovery of the Climate – A New Water Paradigm*. Municipality of Tory.

Krishna, Rama; Kummitha, Reddy; Crutzen, Nathalie. (2017). How Do We Understand Smart Cities? An Evolutionary Perspective, *Cities*, Vol. 67, pp. 43–52.

Kunstler, J.H. (Sept. 1996). *Home From Nowhere*. The Atlantic Monthly.

Landsberg H. The Urban Climate. New York, USA: Academic Press; 1981.

*Landscape architectural review*, 12: 10–12 (1991).

LANGDON, P. *A better place to live: reshaping the American suburb*. Amherst, University ofMassachusetts Press, 1994.

Larsen L., Rajkovich N., Leighton C., McCoy K., Calhoun K, Mallen E, et al. Green Building and Climate Resilience. Understanding impacts and preparing for changing conditions. University of Michigan; U.S. Green Building Council; 2011. 260 p.

LAWRENCE, R. Personal communication. December 8, 1998.

LAWRENCE, R. Wanted: Designs for health in the urban environment. *World health forum*, 17: 363– 366 (1996).

Lawson, B. (1997) How Designers Think: The Design Process Demystified, Oxford Architectural Press: Oxford, UK.

LeCorbusier. La Ville Radieuse: element d'une doctrine d'urbanisme pour l'equipment de la Civilization Machinist. De L'Architecture D'Aujourd'Hui. 1935;

Leung, Y. K., Yeung, K. H., Ginn, E. W. L., and Leung, W. M. (2004).

Levy, J. (2000) Contemporary Urban Planning, Englewood Cliffs, NJ: Prentice Hall.

LiddellH.MackieD.2002Forward to the past. Proceedings of International Conference on Sustainable Building, Oslo, Norway

Lindau, L.A., Hidalgo, D., and Facchini, D. (2010). Curitiba, the Cradle of Bus Rapid Transit. Build Environment (1978). 35(3):274–282.

LINDHEIM, R. & SYME, L. Environments, people and health. *Annual review of public health*, 4: 335–359 (1983).

Liverpool Healthy Cities. *City health plan*. http://dialspace.dial.pipex.com/town/street/ap93/health.htm(1995) (retrieved in October 1998).

Los S. CITTA' SOLARI dal passato al futuro. IUAV. (42):1-16.

Lowry, W. P. (1977). Empirical estimation of urban effects on climate: A problem analysis. *Journal of Applied Meteorology* 16(2), 129–135.

Luhmann N. (2006). La Confiance un mécanisme de réduction de la complexité sociale., 2–111.

LUKE, J. *Catalytic leadership: strategies for an interconnected world*. San Francisco, Jossey-BassPublishers, 1998.

LynchK.1960The image of the cityMIT Press, Cambridge, MA, USA

Magolda, M. (1992) Knowledge and Reasoning in College: Gender-related Patterns in Students Intellectual Development, San Francisco, CA: Jossey-Bass.

Majalah Tempo, (June 2018) PBB: Jumlah Penduduk Dunia 9,8 Miliar Tahun 2050, pp. 18–20.

Marsal-Llacuna, Maria-Lluïsa; Colomer-Llinàs, Joan; Meléndez-Frigola, Joaquim. (2015). Lessons in Urban Monitoring Taken from Sustainable and Livable Cities to Better Address the Smart Cities Initiative. Technological Forecasting and Social Change, pp. 611-622.

MARTI-COSTA, S. & SERRANO-GARCIA, I. Needs assessment and community development: an ideological perspective. *Prevention in human services*, Summer: 75–88 (1983).

Mathison, S. and Freeman, M. (1997) 'The logic of interdisciplinary studies', Paper presented at the Annual Meeting of the American Educational Research Association, Chicago, accessed at http://cela.albany.edu/reports/mathisonlogic12004.pdf, 06-01-2007.

MAXWELL, B. & JACOBSON, M. *Marketing disease to hispanics*. Washington, DC, Center for Sciencein the Public Interest, 1989.

McGranahan, G., et al. (2007). The rising tide: Assessing the risks of climate change and human settlements in low elevation coastal zones. *Environment and Urbanisation* 19, 17–37.

MCKEOWN, T. *Medicine in modern society*. London, Allen & Unwin, 1965.

MCKNIGHT, J. Two tools for well-being. *In:* Minkler, M., ed. *Community organizing and communitybuilding for health*. New Brunswick, Rutgers University Press, 1997, pp. 20–25.

MCMICHAEL, A.J. ET AL., eds. *Climate change and human health*. Geneva, World Health Organization 1996.

*Medical history and medical care*. London, Oxford University Press, 1971, pp. 1–23.

Meijer, Albert; Pedro, Manuel. (2016). Governing the Smart City: A Review of the Literature on Smart Urban Governance, *International Review of Administrative Science*, Vol.82, No.2, pp. 392–408. DOI: 10.1177/0020852314564308.

Mentens, J., Raes, D., and Hemy, M. (2006). Green roofs as a tool for solving the rainwater runoff problem in the urbanized 21st century? *Landscape and Urban Planning* 77, 217–226.

MERCHANT, C. *Radical ecology: the search for a livable world*. New York, Routledge, 1992.

Mikellides, B. (1980) Architecture for People, Austin, TX: Holt, Rinehart and Winston. Muir, T. and Rance, B. (1995) Collaborative Practices in the Built Environment, London, UK: E&FN Spons.

Miller, R. B., and Small, C. (2003). Cities from space: Potential applications of remote sensing in urban environmental research and policy. *Environmental Science & Policy* 6(2), 129–137.

Mills, G. (2005). Urban form, function and climate. Dept. of Geography, University of California, Davis. Accessed August 30, 2014: epa.gov/ heatisland/resources/pdf/ GMills4.pdf

MINKLER, M. Ten commitments for community health education. *Health education research,* 9: 527–534 (1994).

MINKLER, M., S.P. ET AL. The political economy of health: a useful theoretical tool for the health education practice. *International quarterly of health education,* 15: 111–125 (1995).

Mitchell, W. (1999). *e-Topia: urban life, Jim but not as you know it,* Cambridge, MIT Press Molinaro, J. (2014). *National Realtors' Survey Indicates Strong Interest in Walkable, Mixed-Use Neighborhoods.*

MOELLER, D. *Environmental health.* Cambridge, Harvard University Press, 1997.

Mohammad hasan, sharbati, Study of social processes in urban life, with emphasis on cultural and social activities of Mashhad Urban Management.

Montavon M. Optimisation of Urban Form by Evaluation of the Solar Potential. Ecole Polytechnique Federale, Lausanne, France; 2010.

Moonen P, Defraeye T, Dorer V, Blocken B, Carmeliet J. Urban Physics: Effect of the micro-climate on comfort, health and energy demand. Front Archit Res. 2012;1(3):197-228.

Morgan, C., Bevington, C., Levin, D., Robinson, P., Davis, P., Abbott, J., and Simkins, P. (2013). *Water Sensitive Urban Design in the UK – Ideas for Built Environment Practitioners.* Ciria.

Mortimer, T. Jeylan and G. Roberta, Simmons, 1978. Adult Socialization. *Annual Review of Sociology,* 4: 421-54

MUMFORD, L. *The city in history. its origins, its transformations and its prospects.* New York,Harcourt, Brace and World Inc., 1961.

Nathali, B.; Khan, M.; Han, K. (2018). Towards Sustainable Smart Cities: A Review of Trends, Architectures, Components, and Open Challenges in Smart Cities, *Sustainable Cities and Society,* Vol. 38, pp. 697–713.

New York: Princeton Architectural Press. Llewelyn-Davies (in association with Alan Baxter and

Newell, W. (1994) 'Designing interdisciplinary courses', Interdisciplinary Studies Today, Vol. 58, pp.35-51.

Newman, P., & Hogan, T. (1981). A Review of Urban Density Models: Toward a Resolution of the Conflict between Populace and Planner. *Human Ecology, 9*(3), 269-303.

Ng E, Yuan C, Fung JC, Ren C, Chen L. Improving the wind environment in high-density cities by understanding urban morphology and surface roughness: A study in Hong Kong. Landsc Urban Plan. 2011;101(1):59-74.

Ng E. Designin for Urban Ventilation. In: Ng E, editor. Designing High Density Cities - For Social and Environmental Sustainability. London, UK: Earthscan Publications Ltd.; 2009. p. 119-36.

Nikolopoulou M, Baker N, Steemers K. THERMAL COMFORT IN OUTDOOR URBAN SPACES: UNDERSTANDING THE HUMAN PARAMETER. Sol Energy. 2001;70(3):227-35

Nissany, M. (1995) 'Fruits, salads and smoothies: A working definition of interdisciplinarity', Journal of Educational Thought, Vol. 29, No. 2, pp.121-128.

Nowotny, H. (2007) 'The potential of transdisciplinarity', Interdisciplines, www.interdisciplines.org/interdisciplinarity/papers/5/24#_24, accessed May 2, 2007.

Oke, Tim R. (1981). Canyon geometry and the nocturnal urban heat island: Comparison of scale model and field observations. *Journal of Climatology* 1 (3), 237–254.

Oleson, K. W., Monaghan, A., Wilhelmi, O., Barlage, M., Brunsell, N., Feddema, J., Hu, L., Steinhoff, D. F. (2015). Interactions between urbanization, heat stress, and climate change. Climatic Change. 129:525–541.

Olgyay V. Design with Climate: Bioclimatic Approach to Architectural Regionalism. Princeton, USA: Princeton University Press; 1963.

Ong, B. L. (2003). Green plot ratio: An ecological measure for architecture and urban planning. *Landscape and Urban Planning* 63(4), 197–211.

Ottawa charter for health promotion. *Health promotion*, 4: iii–v (1986).

Parolek, D., Parolek, K., Crawford, P. (2008). *Form- Based Codes: A Guide for Planners, Urban Designers Municipalities and Developers*, John Wiley & Sons, Inc., Hoboken, NJ.

Paul, M. J., and Meyer, J. L. (2001). Streams in the urban landscape. *Annual Review of Ecological Systems* 32, 333–365.

Peel, M. C. Finlayson, B. L. , and McMahon, T. A. (2007). Updated world map of the Köppen-Geiger climate classification. *Hydrology and Earth System Sciences Discussions* 4(2), 462.

Peel, M. C., Finlayson, B. L., and McMahon, T. A. 2007. Updated world map of the Köppen-Geiger climate classification. *Hydrology and Earth System Sciences Discussions*, 4 (2), 462.

Perez-Gomez, A. (1983) Architecture and the Crisis of Modern Science, Cambridge, MA: MIT Press.

Petrie, H. (1992) 'Interdisciplinary education: are we faced with insurmountable opportunities?', Review of Research and Education, Vol. 18, pp.299-333.

Picone A. Lo sviluppo sostenibile dell'Oasi di Siwa, un processo corale. Costruire sostenible Il Mediterraneo. Alinea Editrice; 2001. p. 94-9.

Plessner H. Die Sonnenbaulehre des Dr. Bernhardt Christoph Faust: Ein Beitrag zur Geschichte der Hygiene des Staàdtebaus. Technische Hochschule, Berlin, Germany; 1933.

Poleto, C., and Tassi, R. (2012). Sustainable urban drainage systems. In Javaid, M. S. (ed.), *Drainage Systems* (55–72). InTech.

Radeloff, V. C., Hammer, R. B., Stewart, S. I., Fried, J. S., Holcomb, S. S., & McKeefry, J. F. (2005). The Wildland-Urban Interface in the United States. *Ecological Applications, 15*(3), 799-805.

Ratti C, Baker N, Steemers K. Energy consumption and urban texture. Energy Build. 2005;37(7):762-76.

Raven, J. (2011). Cooling the public realm: Climate-resilient urban design · resilient cities. In Otto-Zimmermann, K. (ed.), *Cities and Adaptation to Climate Change: Local Sustainability* (Vol. 1, 451–463), Springer.

Resch, E., Bohne, R.A., Kvamsdal, T., and Lohne, J. (2016). Impact of urban density and building height on energy use in cities. Energy Procedia. 96:800–814.

Rey AA, Pidoux J, Barde C. La Science des Plans de Villes. Paris, France: © Dunod; 1928.

Riding, R. and Cheema, I. (1991) 'Cognitive styles: an overview and integration', Educational Psychology, Vol. 11, pp.193-215.

Rinehart, J. (1999) 'Turning theory into theorizing: collaborative learning in a sociological theory course', Teaching Sociology, Vol. 27, pp.216-232.

Roberts, A. (2005) 'Cognitive styles and student progression in architectural design education', Design Studies, Vol. 27, pp.167-181.

Roth, M. (2007). Review of urban climate research in (sub) tropical regions. *International Journal of Climatology* 27(14), 1859–1873.

Rudosfky B. Architecture without Architects. New York, USA: MOMA: New York; 1964.

Sailor, D. J. (2011). A review of methods for estimating anthropogenic heat and moisture emissions in the urban environment. *International Journal of Climatology* 31(2), 189–199.

Santamouris, M. (2014). Cooling the cities – A review of reflective and green roof mitigation technologies to fight heat island and improve comfort in urban environments. *Solar Energy* 103, 682–703.

Sassen, S. (1991) (2nd ed. 2002). *The Global City*, Princeton, NJ., Princeton University Press.

Scholz, M., and Grabowiecki, P. (2007). Review of permeable pavement systems. *Building and Environment* 42(11), 3830–3836.

Schon, D. (1983) The Reflective Practitioner, New York, NY: Basic Books.

Schuler, M. (2009). The Masdar Development – showcase with global effect.

Scott, J. (1998) Seeing Like a State: How Certain Schemes to Improve the Human Condition Have Failed, New Haven, CT: Yale.

Sebastian, R. (2003) 'Multi-architect design collaboration on integrated urban complex development in the Netherlands', Journal of Design Research, Vol. 3, No. 1.

Secchi B. La Città Normale. In: Mittner D, editor. La città reticolare e il progetto moderno. CittàStudiEdizioni; 2008. p. 47-58.

Shane, D. G. (2005). *Recombinant Urbanism: Conceptual Modeling in Architecture, Urban Design, and City Theory.*

Shashua-Bar, L., Potchter, O., Bitan, A., Boltansky, D., and Yaakov, Y. (2010). Microclimate modelling of street tree species effects within the varied urban morphology in the Mediterranean city of Tel Aviv, Israel. *International Journal of Climatology* 30(1), 44–57.

Silvi C. Solar Building Practices and Urban Planning in the Work of Gaetano Vinaccia (1889-1971). Proceedings of the 2nd International Solar Cities Congress. Oxford, UK; 2006.

Simpson, D. (1998) 'Thinking about educator preparation in the twenty-first century: a Deweyan perspective', Teacher Education Quarterly, Vol. 25, No. 4, pp.96-101.

Sinkonde, K. Daniel. (2018). Modern approach for intelligent database to support urban city accessibility tools for the pedestrian, *Journal of Smart city*, Vol.4, No. 2, pp. 001-007. http://doi.org/10.26789/JSC.2018.02.001.

Smith, C., and Levermore, G. (2008). Designing urban spaces and buildings to improve sustainability and quality of life in a warmer world. *Energy Policy*. doi:10.1016/j.enpol.2008.09.011

STATE OF CALIFORNIA DEPARTMENT OF HEALTH SERVICES. *Data summary: homicide among youngmales, by race/ethnicity,* 1990.

Steemers K, Ratti C, Raydan D. Building form and environmental performance: Archetypes, analysis and an arid climate. Energy Build. 2003;35(1):49-59.

Steger, C. (2000) 'Urban design', J. Levy (Ed.) Contemporary Urban Planning, Englewood Cliffs, NJ: Prentice Hall.

Stewart, I. D., and Oke, T. R. (2012). Local climate zones for urban temperature studies. *Bulletin of the American Meteorological Society* 93(12),1879–1900.

Stone, B., Hess, J., and Frumkin, H. (2010). Urban form and extreme heat events: Are sprawling cities more vulnerable to climate change? *Environmental Health Perspectives* 118, 1425–1428.

Stone, L. D. (1983). The Process of Search Planning: Current Approaches and Continuing Problems. *Operations Research, 31*(2), 207-233.

Symes, M., Pauwells, S., 1999). *The diffusion of innovations in urban design: the case of sustainability in the Hulme development guide.* Journal of Urban Design 4(1), 97–117.

Tablada A, De Troyer F, Blocken B, Carmeliet J, Verschure H. On natural ventilation and thermal comfort in compact urban environments - the Old Havana case. Build Environ. 2009;44(9):1943-58.

Taha, H., Konopacki, S., and Gabersek, S. (1999). Impacts of large scale surface modifications on meteorological conditions and energy use: A 10-region modeling study. *Theoretical and Applied Climatology* 62, 175–185.

TESH, S. *Hidden arguments: political ideology and disease prevention policy.* New Brunswick,Rutgers University Press, 1990.

Theobald, D. M., Spies, T., Kline, J., Maxwell, B., Hobbs, N. T., & Dale, V. H. (2005). Ecological Support for Rural Land-Use Planning. *Ecological Applications, 15*(6), 1906-1914.

Thomas, R., and Ritchie, A. (eds.). (2003). *Sustainable Urban Design: An Environmental Approach.* Spon Press.

Trindade, E. Priscila; Hinnig, Marcus PF; Moreira da Costa, Eduardo; Marques, J. Sabatini; Bastos, Rogério C; Yigitcanlar, Tan. (2017). Sustainable Development of Smart Cities: A Systematic Review of the Literature, *Journal of Open Innovation: Technology, Market, and Complexity.* Vol.3, No.11. DOI: 10.1186/s40852-017-0063-2

*Twenty steps for developing a Healthy Cities project,* 3rd ed. Copenhagen, WHO Regional Office for Europe, 1997 (document EUR/ICP/HSC 644(2)).

Urban adaptation can roll back warming of emerging megapolitan regions. *Proceedings of the National Academy of Sciences* 111(8), 2909–2914.

Urbanization signatures in strong versus weak precipitation over the Pearl River Delta metropolitan regions of China. *Environmental Research Letters* 6(3), 034020.

Vegas F, Mileto C, Cristini V, Checa JRR. Underground cities. In: Correia M, Dipasquale L, Mecca S, editors. VERSUS: HERITAGE FOR TOMORROW Vernacular Knowledge for Sustainable Architecture. Florence, Italy: Firenze University Press; 2014. p. 114-27.

Vinaccia G. Il Problema dell'Orientamento nell'Urbanistica dell'Antica Roma. Roma: Istituto di Studi Romani; 1939. 42 p.

Vinaccia G. La Città di Domani. Vol1. Come il Clima Plasma la Forma Urbana e l'Architettura. Roma: Fratelli Palombi Editore; 1943. 155 p.

Vogt A. Über die Richtung der städtischen Straßen nach der Himmelsgegend und das Verhältnis ihrer Breite zur Häuserhöhe. Z Biol. 1879;

Von Gerkan A. Griechische Stadteanlagen. Berlin Leipzig; 1924.

Walker, B., and Salt, D. (2006). *Resilience Thinking*. Island Press.

WALLACE, D. Smaller increases in life expectancy for blacks and whites between the 1970s and1980s. *American journal of public health*, 85: 875–876 (1990).

WALLACE, R. & WALLACE, D. Community marginalisation and the diffusion of disease and disorderin the United States. *British medical journal*, 314: 1341–1345 (1997).

WALLACK, L. *Introduction to public health*. Lecture presented in Public Health 200 1997. School ofPublic Health, University of California, Berkeley, CA, 1997.

Walters, D. (2007). *Designing Community: Charrettes, Masterplans and Form-based Codes*. Architectural Press, Oxford, UK.

Walters, D. (2011). *Smart Cities, Smart Places, Smart Democracy: Form-based codes, electronic governance and the role of place in making smart cities*. Intelligent Buildings International, 3, 198-218.

Walters, D. and Brown, L. (2004). *Design First: Designbased Planning for Communities*. Oxford: Architectural Press.

Walters, D., and Brown, L. (2004). *Design First: Designbased Planning for Communities*, Oxford, Architectural Press.

Walters, D., Brown, L. (2004). *Design First: Design-based Planning*

WANDERSMAN, A. ET AL. Understanding coalitions and how they operate: an 'open systems' organizational framework. *In*: Minkler, M. ed. *Community organizing and community building for health*. New Brunswick, Rutgers University Press, 1997, pp 261–277.

WANG, S. & SMITH, P.J. In quest of forgiving environment: residential planning and pedestrian safetyin Edmonton, Canada. *Planning perspective*, 12: 225–250 (1997).

Wang, Z.-b. (2009, 20-22 Sept. 2009). *Review and Prospect of Urban Planning Management Information System in China.* Paper presented at the Management and Service Science, 2009. MASS '09. International Conference on.

Warburton, D., and Yoshimura, S. (2005). Local to global connections. In Velasquez, J., Yashiro, M., Yoshimura, S., and Ono, I. (eds.), *Innovative Communities: People-centred Approaches to Environmental Management in the Asia-Pacific Region,* United Nations University Press.

Ward, J. (2004) 'The making of a library', Metropolis, Vol. 24, No. 2, pp.97-115.

WEKERLE, G., & WHITZMAN, C. *Safe cities: guidelines for planning, design, and management.* NewYork, Van Nostran Reinhold, 1995.

WHO REGIONAL OFFICE FOR EUROPE. *Health in Europe: the 1993/94 health for all monitoring report.* Copenhagen, WHO Regional Office for Europe, 1994 (WHO Regional Publications, European Series, No. 56).

Wilson, W. (1989) The City Beautiful Movement, Baltimore, MD: Johns Hopkins University.

World Bank. (2017). 2016 GNI per capita, Atlas method (current US$).

WORLD HEALTH ORGANIZATION. Constitution of the World Health Organization. *In: WHO basicdocuments,* 40th ed. Geneva, World Health Organization, 1994.

Wyatt, R. (2004) 'The great divide: difference in style between architects and urban planners', Journal of Architecture and Planning Research, Vol. 21, No. 1, pp.38-53.

Yewlett, C. J. L. (2001). OR in Strategic Land-Use Planning. *The Journal of the Operational Research Society, 52*(1), 4-13.

Yinghui, X., & Qingming, Z. (2009, 20-22 May 2009). *A review of remote sensing applications in urban planning and management in China.* Paper presented at the Urban Remote Sensing Event, 2009 Joint.

Zhao, S.X., Guo, N.S., Li, C.L.K., Smith, C. (2017). Megacities, the World's Largest Cities Unleashed: Major Trends and Dynamics in Contemporary Global Urban Development. World Development. 98:257–289.

Zheng, H. and Peeta, S. (2015). Network design for personal rapid transit under transit-oriented development. Transportation Research Part C. 55:351–362.

# INDEX

www.ingramcontent.com/pod-product-compliance
Lightning Source LLC
Chambersburg PA
CBHW062003190326
41458CB00009B/2952